AN INTRODUCTION TO THE SEISMICITY OF THE UNITED STATES

Monograph Series

Engineering Monographs on Earthquake Criteria, Structural Design, and Strong Motion Records

Coordinating Editor, Mihran S. Agbabian

Monographs Available

Reading and Interpreting Strong Motion Accelerograms, by Donald E. Hudson

Dynamics of Structures — A Primer, by Anil K. Chopra

Earthquake Spectra and Design, by Nathan M. Newmark and William J. Hall

Earthquake Design Criteria, by George W. Housner and Paul C. Jennings

Ground Motions and Soil Liquefaction during Earthquakes, by H. Bolton Seed and I.M. Idriss

Seismic Design Codes and Procedures, by Glen V. Berg

An Introduction to the Seismicity of the United States, by S. T. Algermissen

AN INTRODUCTION TO THE SEISMICITY OF THE UNITED STATES

S. T. Algermissen

U.S. Geological Survey
Denver

EARTHQUAKE ENGINEERING RESEARCH INSTITUTE

Published by

The Earthquake Engineering Research Institute, whose objectives are the advancement of the science and practice of earthquake engineering and the solution of national earthquake engineering problems.

This is volume seven of a series titled: Engineering Monographs on Earthquake Criteria, Structural Design, and Strong Motion Records.

The publication of this monograph was supported by a grant from the *National Science Foundation*.

Library of Congress Catalog Card Number 83-80955
ISBN 0-943198-26-7

This monograph may be obtained from:
    Earthquake Engineering Research Institute
    2620 Telegraph Avenue
    Berkeley, California 94704

The views expressed in this monograph are those of the author and do not necessarily represent the views or policies of the Earthquake Engineering Research Institute, the National Science Foundation, or the U.S. Geological Survey.

Dedicated to

Marta and Carl Andrew

FOREWORD

This monograph by S. T. Algermissen on the seismicity of the United States is the seventh in a series of monographs on different aspects of earthquake engineering. The monographs are by experts especially qualified to prepare expositions of the subjects. Each monograph covers a single topic, with more thorough treatment than would be given to it in a textbook on earthquake engineering. The monograph series grew out of the seminars on earthquake engineering that were organized by the Earthquake Engineering Research Institute and presented to some 2,000 engineers. The seminars were given in Los Angeles, San Francisco, Chicago, Washington D.C., Seattle, St. Louis, Mayaguez P.R., and Houston. The seminars were aimed at acquainting engineers, building officials and members of government agencies with the basics of earthquake engineering. In the course of these seminars it became apparent that a more detailed written presentation would be of value to those wishing to study carthquake engineering, and this led to the monograph project. The present monograph discusses the seismicity of the United States including Alaska and Hawaii and also Puerto Rico. Seismic zones and the development of national seismicity maps are also presented. These topics are of interest not only to engineers and building planners, but to all residents of the United States and Puerto Rico.

The EERI monograph project, and also the seminar series, was supported by the National Science Foundation. EERI member M. S. Agbabian served as Coordinator of the seminar series and is Coordinating Editor of the monograph project. Technical editor for the series is J. W. Athey. Each monograph is reviewed by the members of the Monograph Committee: M.S. Agbabian,

G.V. Berg, R.W. Clough, H.J. Degenkolb, G.W. Housner, and C.W. Pinkham, with the objective of maintaining a high standard of presentation.

Monograph Committee
November, 1983

PREFACE

This volume provides an introduction to an understanding of the seismicity of the United States. Seismological research in the United States has greatly expanded in the past 20 years. The great Alaska earthquake of 1964, a realization of the importance of the siting of critical facilities and the implementation of a worldwide network of seismograph stations, are only three of a number of reasons for the expansion of both research in seismology and the application of the research to seismic zoning and related problems. There is, of course, an extensive literature on seismicity and seismic zoning but much of this literature is dispersed in a variety of journals and books, making it difficult for the non-specialist and general reader to find a summary of the subject that provides an overall view of the characteristics of earthquakes throughout the diverse geological environments that make up the country. It is my hope that this monograph will provide some useful information to the non-specialist.

References are provided for those who wish to study particular regions or subjects in more detail. It would be very difficult in a work of this type to provide a comprehensive list of references, but my aim has been to at least provide sources of additional reading that are useful and informative.

The seismicity of the country is discussed geographically, and it may appear that areas of the country with low to moderate seismicity such as the Midwest and East have received a disproportionate amount of coverage with respect to areas such as California-Nevada and Alaska. This was done for two reasons: first, the general reader is probably more familiar with the seismological literature for the areas of the country with a high hazard and, second, a large earthquake, even though one occurs infrequently in an area of moderate seismicity, may result in very disastrous losses. The general population is not very aware of the earthquake problem in areas of moderate seismicity, and few if any structures have been tested by earthquake-induced shaking. The chapter on zoning provides background information about the application of seismological and geological research to the problems of earthquake-resistant design.

S. T. Algermissen
August, 1983

TABLE OF CONTENTS

	PAGE
Introduction	13
Northeast Region	25
Southeast Region	30
Central Region	39
Western Mountain Region	54
Pacific Region	65
California and Western Nevada	65
Washington and Oregon	77
Alaska	85
Hawaii	95
Puerto Rico	99
Seismic Zoning in the United States	104
Introduction	104
Development of National Maps	105
Acknowledgements	119
Appendix A: Modified Mercalli Intensity Scale	120
Appendix B: Magnitude	127
References	130

An Introduction to the Seismicity of the United States

by S. T. Algermissen

INTRODUCTION

The assessment of earthquake hazards depends upon a great many factors, such as: (1) the length and completeness of the historical record of seismicity; (2) geological evidence of earthquake occurrence and recognition of local or regional geological structures that may indicate a potential for the occurrence of damaging earthquakes even though little or no historical seismicity is known for the area; (3) the nature of seismic source characteristics, seismic wave propagation and the effects of local site conditions on seismic waves; and (4) geologic situations that may result in ground failures such as liquefaction and landsliding.

The historical seismic record has been widely used as a basis for the estimation of seismic hazard because of the relatively small amount of quantitative geological evidence of earthquake occurrence (for example, fault slip) available and the restricted geographical distribution of these data. While a great amount of additional geological data has emerged in the past ten years, the historical record of earthquake occurrence still forms the principal basis for estimating the earthquake hazard of much of the country.

The known historical record of earthquakes in the United States is relatively short. The only data available for most earthquakes prior to about 1900 are historical accounts of earthquake effects. These accounts have been used to estimate the distribution of intensities and, where possible, the maximum intensity of the earthquakes. The intensities are derived from observations of the extent of human and animal reaction to ground shaking, the effects of shaking on structures, trees, bushes, etc. and geological effects such as landsliding and liquefaction. The perception and observation of earthquake effects are summarized and correlated with written descriptions of degrees of shaking contained in an intensity scale. Since 1931 the Modified Mercalli Intensity (MMI) Scale has been widely used in the United States (Ref. 1). Prior to 1931, an intensity scale developed by Rossi and Forel was used (Ref. 2) but in most modern earthquake catalogs for the United States, Rossi-Forel intensities have been converted to Modified Mercalli intensities.

The assignment of intensities has a number of difficulties associated with it. Ideally, intensity values are assigned on the basis of the effects at a particular location or at most over a small area. In practice, they have often been generalized over fairly large areas. Intensity assignment is dependent upon the distribution of population and, for older earthquakes, the availability of media for publication of earthquake descriptions (such as newspapers) and communication (such as the telegraph, telephone, etc.). Thus, for many older earthquakes the only known effects are those reported from cities large enough to have had newspapers. In many cases only an average intensity can be assigned for a rather large city or large area. Despite the subjective nature of the MMI scale, intensity data provide much information about the size and distribution of effects associated with earthquakes that cannot be obtained in any other manner. Intensity data provide invaluable information on attenuation of ground motion in areas such as the Midwest and eastern United States where strong ground motion records are sparse. The problem with intensity observations is that they cannot be easily correlated with more quantitative measures of strong ground motion such as peak acceleration, peak velocity, spectral values, etc. Recent studies, however, indicate that statistical correlations relating response spectral amplitudes of strong ground motion to intensity over select period bands in some instances show lower levels of statistical uncertainty than correlations among peak ground motion values and intensity (Ref. 3). Perhaps additional work will extend the usefulness of intensity data. For a description of the Modified Mercalli Scale, see Appendix A.

The eastern United States generally has a longer complete

record of earthquakes because of its earlier and generally more uniform population distribution than the western states, Alaska and Hawaii. Hawaii presents special problems because of its short historical record of earthquakes and because of the difficulties in locating offshore shocks. The East and Midwest have an additional advantage with respect to the completeness of the earthquake record, for their earthquakes are felt over areas 10-30 times larger than earthquakes of equal magnitude in California and much of the West (Ref. 4). Thus, earthquakes in the East and Midwest are much more likely to be reported than in other areas.

The seismicity of the United States can be depicted in a number of different ways. A useful overview for engineering purposes is provided by examination of the location and maximum Modified Mercalli intensity of damaging earthquakes throughout the country. Figure 1 depicts the spatial distribution of earthquakes with maximum Modified Mercalli intensities of V or greater known to have occurred in the United States and Puerto Rico through 1976. Intensity V is the degree of the MMI scale at which very minor damage such as cracking of plaster will occasionally occur. Figure 1 is useful in an engineering sense because it maps the historical seismicity of the country in terms of damage. While the MMI scale is unsatisfactory in many ways as a quantitative measure of earthquake damage, it represents practically the only measure of the size and effects of an earthquake that can be used over the entire historical record of earthquakes in the United States.

The record of earthquakes in the United States is believed to have started with the Rhode Island earthquake of 1568 (Ref. 5). The history of large earthquakes in the 19th century — those with maximum Modified Mercalli intensities of VIII and greater — is reasonably well documented for the eastern United States but not for other parts of the country. The earliest documented shock in California was in 1769 (Ref. 2). Instrumental data from stations in the United States were not available until after 1887 (Ref. 6) when the first seismograph stations in the country were established at Berkeley and Mt. Hamilton (Lick Observatory). The earthquake catalog of the University of California at

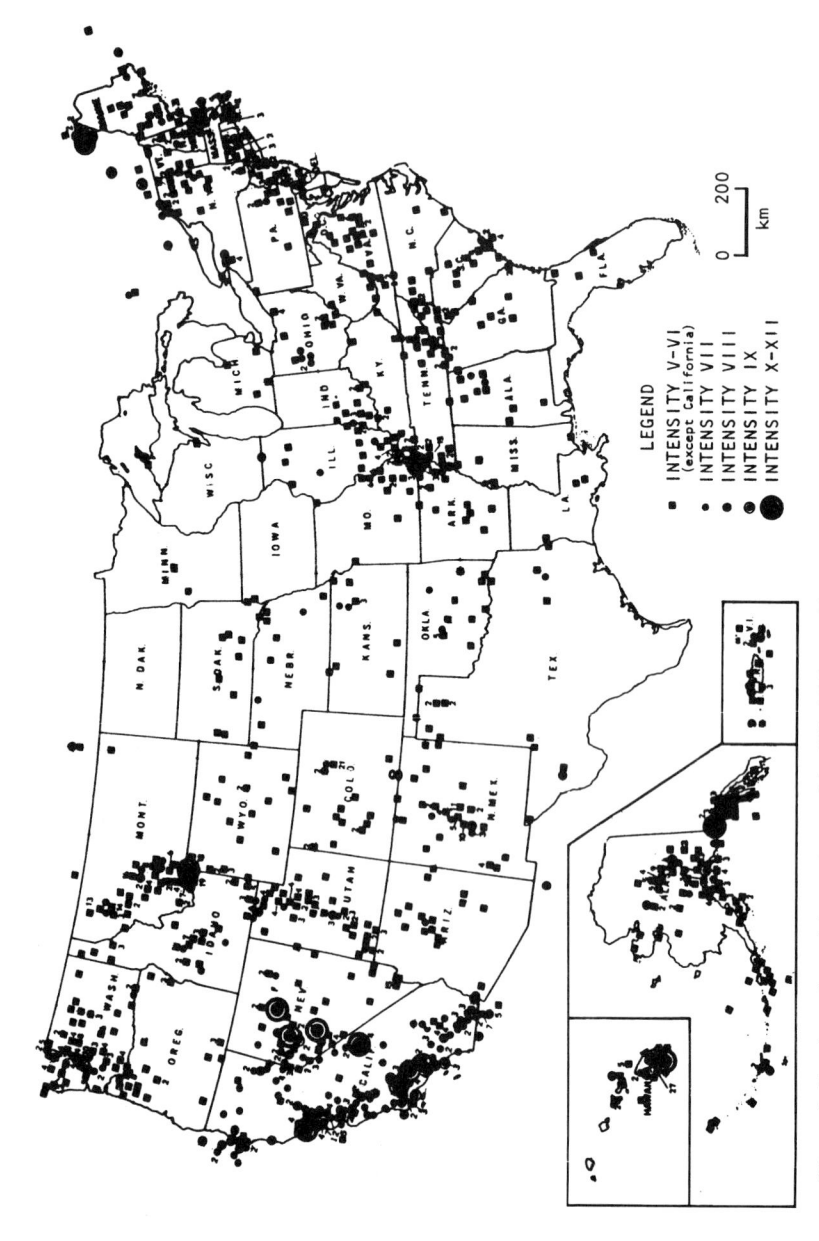

Figure 1. Earthquakes with maximum Modified Mercalli intensities of V or above in the United States and Puerto Rico through 1976 (Ref. 18 with some modifications).

Berkeley did not begin until 1910. The catalog at California Institute of Technology was first published in 1932. Records of earthquakes with magnitudes (local magnitudes in most cases) above 5 are believed to be incomplete for the western part of the country until the early to mid 1960's (Ref. 7). See Appendix B for a discussion of earthquake magnitude and the relationships between various magnitudes.

An estimate of the detection capability of seismograph stations in the conterminous United States in 1965 is shown in Fig. 2. The magnitude (m_b) level at which at least five stations (of the existing network of about 104 stations at that time) should detect the earthquake is shown. Detection by five stations is needed because specification of the hypocenter of an earthquake requires determination of four quantities: longitude, latitude, depth of focus and origin time of the earthquake (so that at least four arrival times of earthquake waves are required, and most routine computer programs for hypocenter determination are designed to operate with not fewer than five observations). The magnitude threshold shown in Fig. 2 assumes that: (1) all stations in existence in 1965 are operating all of the time and (2) all of the observations from the stations are collected and available for hypocenter computations. Clearly, these assumptions are not met most of the time. Seismograph stations in the United States are maintained and operated by such diverse organizations as government agencies, universities, private companies and individuals. Only a few groups or organizations actually compute earthquake hypocenters. If seismograph station data are not routinely used in earthquake hypocenter determination, and the resulting hypocenters published in the seismological literature, the station data will have little or no impact on earthquake hazard evaluation. Figure 3 shows how the magnitude detection threshold decreased in a portion of the Midwest between 1970 and 1980. The dramatic decrease in magnitude detection threshold in this ten-year period is primarily the result of the installation of rather dense networks of stations to study known earthquake source zones near Anna, Ohio and in the Mississippi River Valley (Ref. 9). The magnitude detection threshold has also decreased greatly, but in a very irregular fashion for much of the

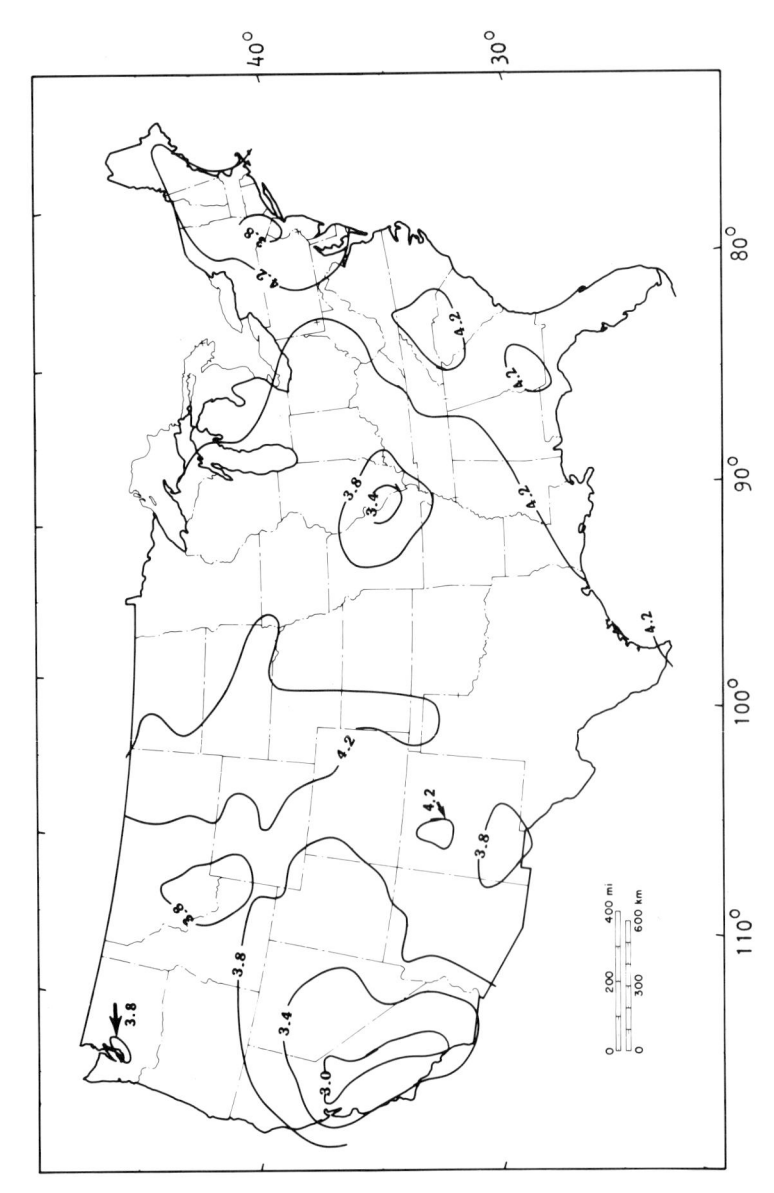

Figure 2. Approximate minimum magnitude (m_b) level at which an earthquake would be detected by at least five stations, given the National Network of about 104 seismograph stations operating regularly in 1965 (Ref. 8). The magnitude level for detection by five or more stations is much lower for most of the country at the present time.

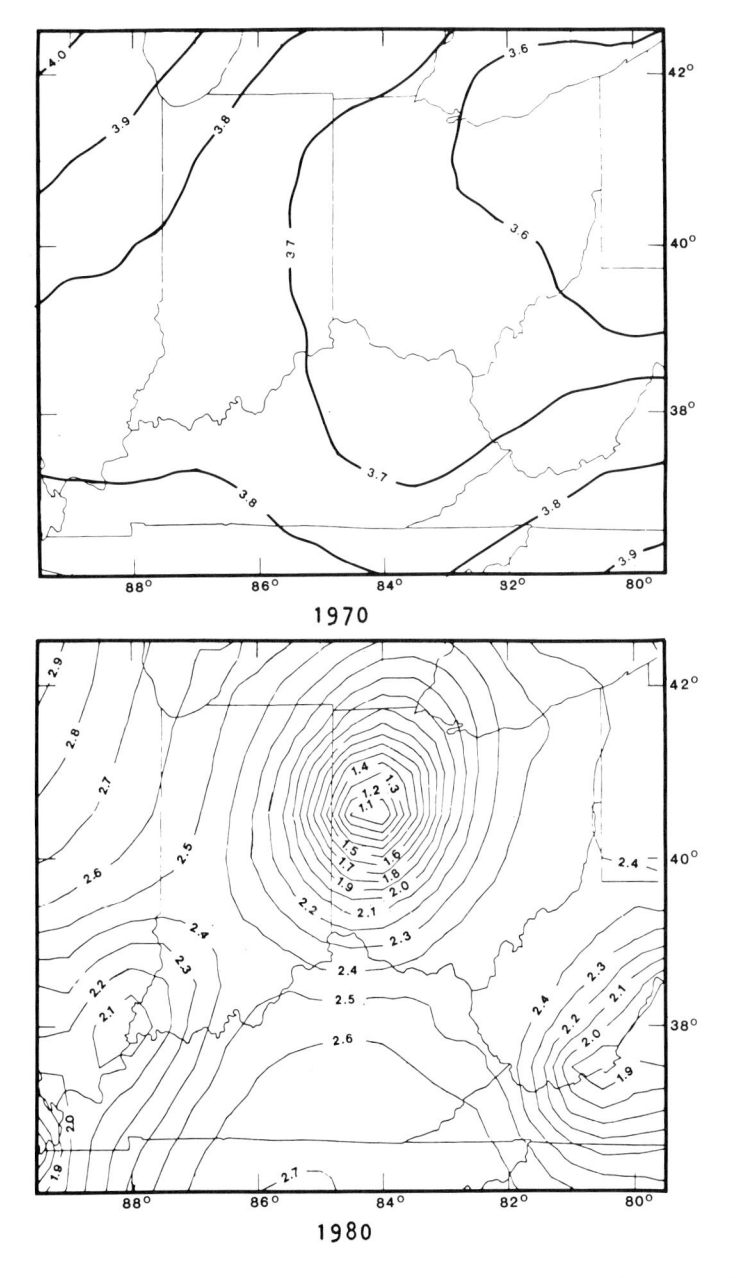

Figure 3. Comparison of five-station detection threshold magnitude (m_b) in 1970 and in 1980 for a portion of the Midwest (Ref. 9).

country and especially in California. Unfortunately, regional networks, particularly in the Midwest and East, are often operated for only a few years and then discontinued.

Earthquakes that occur near or beneath the oceans bordering the United States cannot be well located using only seismograph station data from within the country. In addition, stations within the country do not provide an adequate network of data for the determination of other useful parameters of earthquakes such as instrumentally determined focal mechanisms. An important advance in national and international seismology took place between 1960 and 1967 when 120 seismograph stations (known as the Worldwide Network of Standard Seismograph Stations, WWNSS) were installed throughout the world by the U.S. Coast and Geodetic Survey with funding from the U.S. Department of Defense. The stations have standardized three component short and long period seismographs and crystal clocks accurate to 1 part in 10^7 with provision for reception and recording of standard radio time signals (Ref. 10). The stations currently operating in the network are shown in Fig. 4. Portions of this international network have been periodically updated with improved instrumentation, and additional modern stations have been installed under similar programs (Ref. 11). At the present time, seismological data are available on a routine basis to the National Earthquake Information Service (NEIS) of the U.S. Geological Survey from over 400 seismograph stations outside the United States. Reference 6 lists station identification codes for 2,266 seismograph stations that have operated for some period of time or that are currently in operation in the United States. Many of these stations are no longer in operation but additional new stations are constantly being opened. The NEIS annually computes hypocenters for about 6000 earthquakes worldwide. The hypocenter and related earthquake data are released in a variety of publications (Refs. 12, 13, 14, 15).

The foregoing discussion only touches upon the complexities involved in making use of the historical seismicity record in hazard evaluation. The constantly changing distribution of seismograph stations of extremely uneven quality together with the variability virtually built in to the non-instrumental record

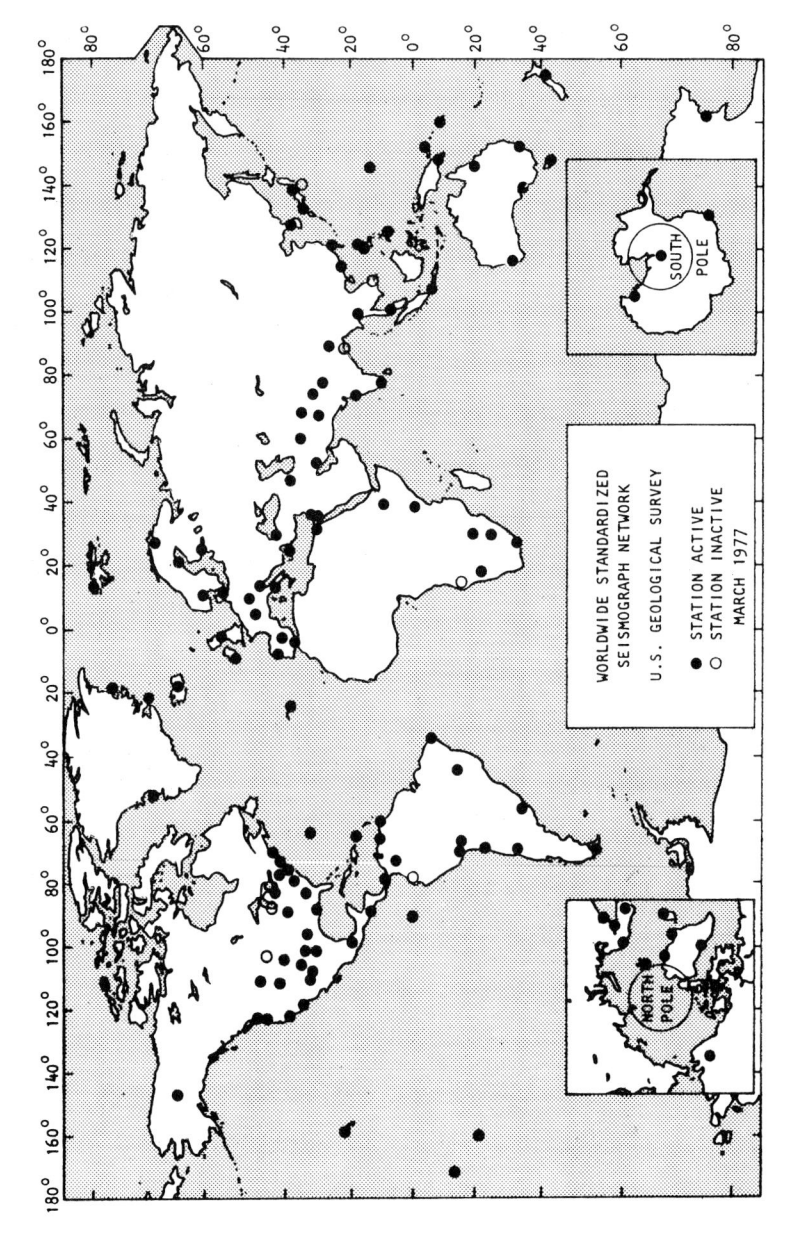

Figure 4. Stations currently in the Worldwide Network of Standard Seismograph Stations (WWNSS).

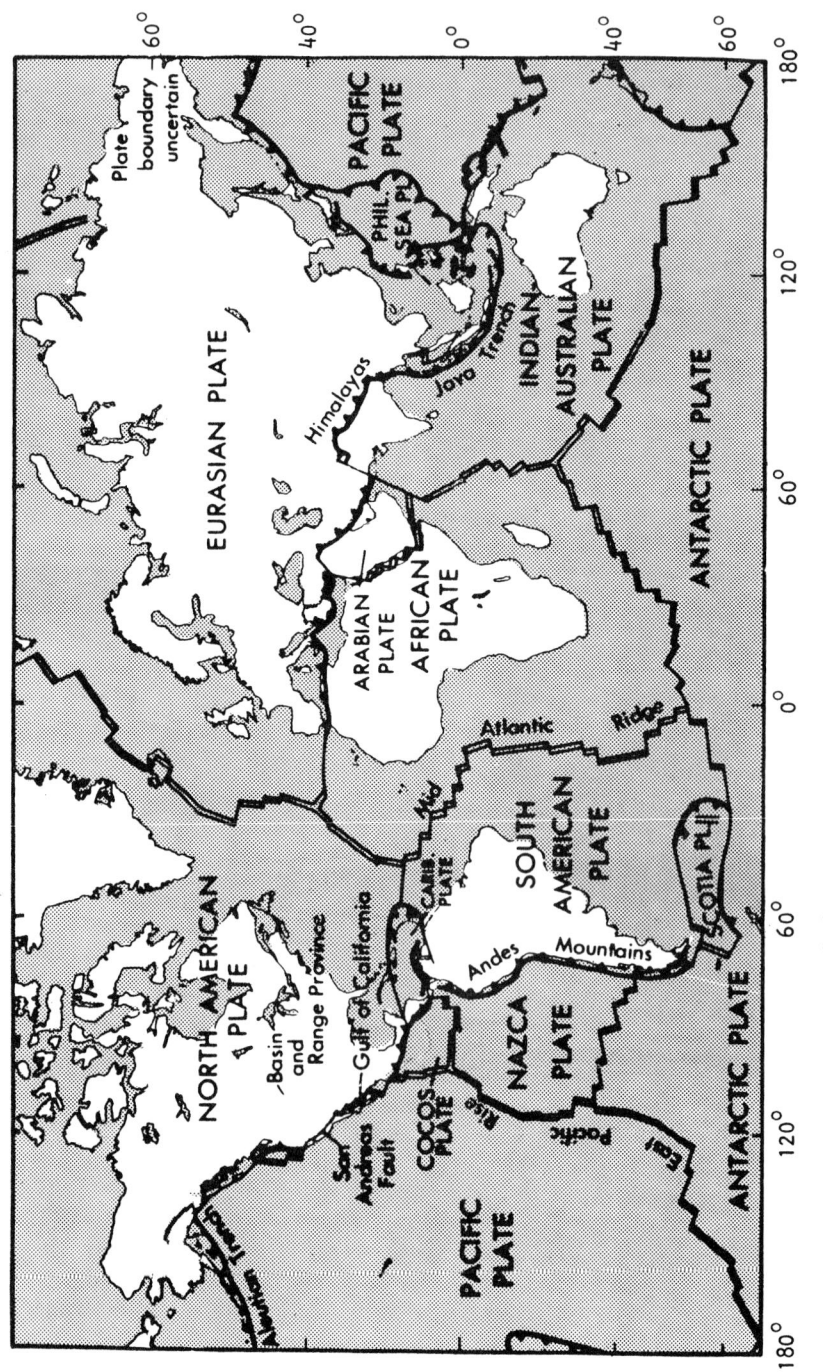

Figure 5. Principal plates, ridges and subduction zones (solid triangles).

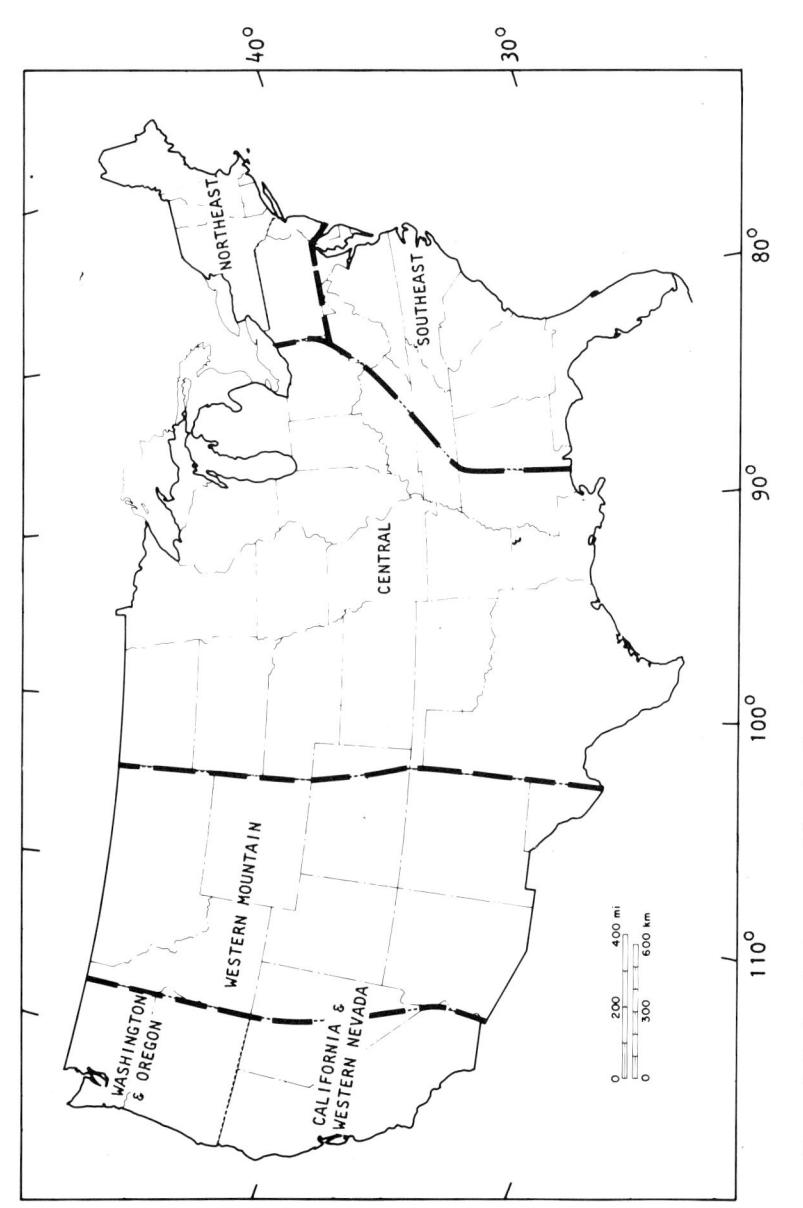

Figure 6. Regional scheme used for the discussion of the seismicity of the conterminous United States.

presents considerable problems in estimating the completeness of the seismic history for any studies of seismic hazard. Intensity is used here in the discussion of the seismicity of the country and, with the exception of Alaska, epicentral maps show the maximum intensity of each earthquake. Magnitudes of significant earthquakes are indicated as appropriate.

The development over the past 70 years of the concept of plate tectonics as a hypothesis and the definition of the hypothesis in a series of papers about 15 years ago provide a framework for the understanding of the worldwide distribution of earthquakes and for earthquake hazard evaluation. Many excellent references both popular and specialized are available (Refs. 16, 17), and thus no general discussion of plate tectonics will be given here. Figure 5 depicts the principal plates, ridges and subduction zones around the world. For the United States, earthquakes associated in various ways with plate boundaries occur in Alaska, the Pacific Northwest, California and the Caribbean. Intraplate earthquakes occur throughout the contiguous United States. The relationships between tectonic processes and the occurrence of earthquakes within the North American plate, particularly in the Midwest and eastern United States, are not well understood even though current research has provided many promising ideas.

The plan of this book is to discuss, on a regional basis, the seismicity of the United States and some ideas about the relationship of the seismicity to geologic structure. The conterminous United States is divided into regions (Fig. 6) following approximately the scheme used in the publication *Earthquake History of the United States* (Ref. 18).

NORTHEAST REGION

Earthquakes with magnitudes of at least 7.0 have occurred in New England and the St. Lawrence River Valley in Canada. The historic seismicity (Refs. 5, 19) is shown in Fig. 7 and the larger shocks of the region are listed in Table 1. The magnitudes (M) listed in Table 1 have been computed using a variety of techniques. The magnitudes are, however, approximately equal to M_s. The maximum intensities for many of the early earthquakes and those offshore have been estimated using whatever intensity data are available. The major concentrations of seismicity in eastern Canada and in the St. Lawrence River Valley have been called the western Quebec, Charlevoix, lower St. Lawrence, and Attica-Niagara zones (Ref. 22). These zones are shown hachured in Fig. 7 as a convenience in describing the seismicity.

Several alignments of earthquake epicenters have been suggested for the Northeast. A northeast-southwest trend along the St. Lawrence Valley to the southeast Missouri (New Madrid) seismic zone was suggested by Woollard (Ref. 23). The entire trend lies in lowlands where the elevations do not exceed 200 meters (Ref. 24). This alignment of epicenters is, however, discontinuous from the St. Lawrence Valley to southeast Missouri, and geologic investigation of the area has yielded little evidence of faulting or other geological structures to support the idea.

A northwest-southeast pattern of seismic activity from the western Quebec area through New England to the coast near Boston, known as the Boston-Ottawa trend, has also been proposed. A major fault system, the Ottawa-Bonnechere graben, is located near the southwest edge but within the western Quebec zone near the southwestern edge of the Boston-Ottawa trend. The projection of the Boston-Ottawa trend offshore into the Atlantic coincides with the strike of the New England (Kelvin) seamount chain, a major topographic and tectonic feature (Refs.

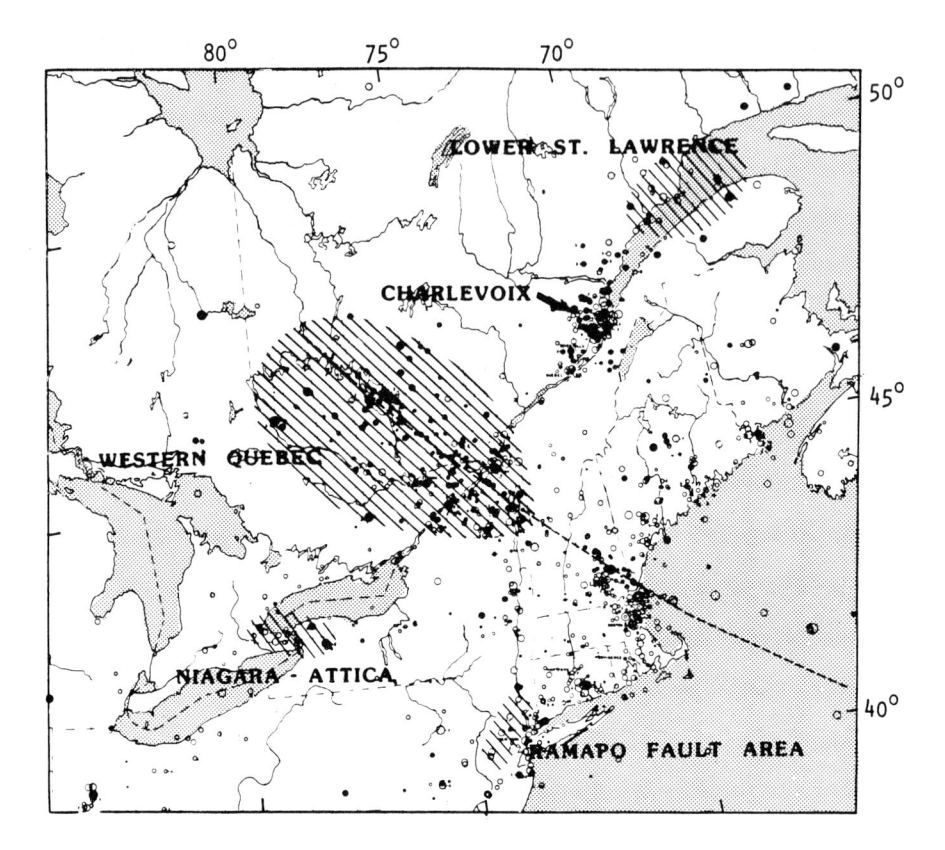

Figure 7. The seismicity of the northeast region of the United States and eastern Canada for the period 1534-1959 (Refs. 5, 19). The solid circles are principally instrumentally determined epicenters, while the open circles represent earthquakes located using intensity data. The hachured and named areas represent concentrations of seismicity grouped together only for the purpose of discussion in the text. The dashed line represents the strike of New England (Kelvin) sea mount chain offshore. Onshore, the line has been extended to show the northwest-southeast alignment of seismicity known as Boston-Ottawa trend.

25, 26, 27). Onshore, the seismicity trend roughly coincides with a zone of Mesozoic alkaline magmatism. The idea of the Boston-Ottawa trend has been challenged on the basis of the generally low seismicity along the trend in Vermont.

Another feature of the seismicity is a rather broad, diffuse trend extending in a north-northeast direction from New Jersey through New England to New Brunswick, Canada (Ref. 28). On January 9, 1982, an earthquake with a magnitude (m_b) of 5.7 occurred in northcentral New Brunswick, followed by several strong aftershocks in the next few days and hundreds of small aftershocks in the following months. No structural damage is known to have occurred (Ref. 21). It has been added to Table 1 because it is the largest shock onshore in northeastern North America since 1944. Considerable seismic activity in the New

Table 1

Important Earthquakes of Eastern Canada and New England

Date (GMT)[‡]	Location	Maximum MM Intensity (I_0)*	Magnitude (M)**
1534-1535	St. Lawrence Valley ?	IX-X	
Jun 11, 1638	St. Lawrence Valley ?	IX	
Feb 5, 1663	Charlevoix Zone	X	7.0
Nov 10, 1727	Near Newbury, MA	VIII	
Sep 16, 1732	Near Montreal	VIII	7.0
Nov 18, 1755	Near Cape Ann, MA	VIII	
May 16, 1791	East Haddam, CT	VIII	
Oct 5, 1817	Woburn, MA	VII-VIII	
Oct 17, 1860	Charlevoix Zone	VIII-IX	6.0
Oct 20, 1870	Charlevoix Zone	IX	6.5
Mar 1, 1925	Charlevoix Zone	IX	7.0
Aug 12, 1929	Attica, NY	VIII	5.5
Nov 18, 1929	Grand Banks of Newfoundland	(X)***	7.0
Nov 1, 1935	Timiskaming, Quebec	VII	6.0
Sep 5, 1944	Massena, NY - Cornwall, Ont.	VIII	6.0
Jan 9, 1982	Northcentral New Brunswick	V	5.7m_b

*All maximum intensities are from Ref. 18 except for the earthquakes of 1732 (Ref. 20), 1935 (Ref. 19) and 1982 (Ref. 21).

**All magnitudes are from Ref. 22 except for the earthquake of 1982 (Ref. 21). Also, see Appendix B for explanation of M.

***Estimated maximum intensity (offshore)

[‡]Greenwich Mean Time

York City area has been found along several northeast trending faults west of New York City in northern New Jersey and southern New York State. The most active of these is the Ramapo fault system which forms the boundary of a Triassic-Jurassic graben known as the Newark graben (Ref. 29 and Fig. 7).

The high concentration of earthquakes in the Boston vicinity (see Fig. 7) may, in part, be related to the early colonization of the area, and hence more reportage. Seismicity in the period 1970-79 indicates that coastal Maine and New Brunswick are, at least currently, as seismically active as the Boston area (Ref. 29). There is also evidence that the seismicity of the eastern Massachusetts-New Hampshire area was higher during the period 1725 through 1824 than in the following 100 years from 1825 through 1924 (Ref. 30). Important earthquake activity has also occurred offshore east of Boston. In 1755, an earthquake located off Cape Ann (maximum MM intensity approximately VIII) caused considerable damage in Boston, was felt from Chesapeake Bay, MD to the Annapolis River, Nova Scotia, and was followed by an aftershock sequence lasting at least several months. The large, November 18, 1929 shock beneath the Grand Banks of Newfoundland was felt throughout New England. This earthquake was responsible for the rupture of twelve trans-Atlantic cables crossing the epicentral area (epicenter

Table 2

Distribution of Earthquakes by Maximum Intensity in the Northeast Region*
(through 1976)

Maximum MM Intensity	Number
V	120
VI	37
VII	10
VIII	3

*Source: Ref. 18 and U.S. Geological Survey earthquake data file. Canadian earthquakes and earthquakes believed to have offshore epicenters are not included.

approximately 44° N lat., 56° W long.). The resulting tsunami (seismic sea wave) caused considerable damage and some life loss at Placentia Bay, Newfoundland. Small tsunamis were recorded along the Atlantic Coast as far south as Charleston, South Carolina.

The distribution of earthquakes with regard to maximum Modified Mercalli intensity in the northeastern region is shown in Table 2. The table includes only earthquakes within the United States, and not Canadian shocks and earthquakes with epicenters believed to be offshore. Note that the three largest earthquakes within the region have maximum Modified Mercalli intensities of VIII.

SOUTHEAST REGION

The southeast region is an area of diffuse, low level seismicity with concentrations of activity both parallel and perpendicular to the generally northeast structural trend of the Appalachians. Regional geological features are shown in Fig. 8, and the seismicity of the region is shown in Fig. 9. Bollinger (Ref. 33) has divided the earthquake activity into four zones:

1. *Southern Appalachian seismic zone.* Zone extending from western Virginia to central Alabama in the Valley and Ridge and Blue Ridge provinces.

2. *Northern Virginia-Maryland seismic zone.* A diffuse northeastward extension of the Southern Appalachian zone.

3. *Central Virginia seismic zone.* A relatively narrow, isolated zone of activity, offset from the above two zones and located in the Piedmont Province; oblique to the NE-SW structural grain.

4. *South Carolina-Georgia seismic zone.* A broad zone spanning both the Piedmont and Coastal Plain provinces; transverse to the regional structure.

General observations can be made about the current rate of seismicity. The seismicity of the area has been lower in recent decades, at least with regard to earthquakes with maximum MMI's of VII and greater. An event with a maximum MMI of VII has not been reported in nearly 60 years, and earthquakes with a maximum MMI of VIII have not occurred in nearly 80 years. Recurrence relationships computed for this area indicate a higher average rate of occurrence of earthquakes with maximum MMI's of VII and VIII than have actually been observed (Ref. 32).

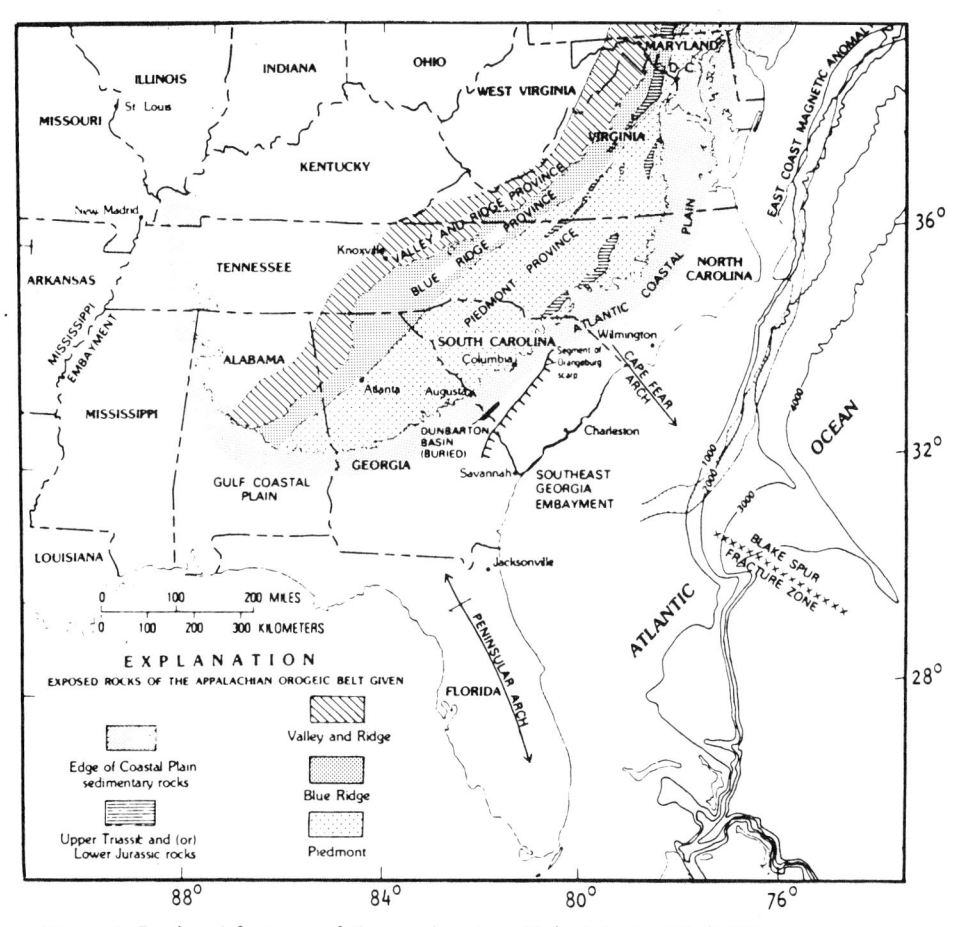

Figure 8. Regional features of the southeastern United States (Ref. 31). The contour interval for submarine topography is 1000 meters.

The largest and by far the most destructive earthquake in the region occurred on August 31, 1886, with its epicenter about 25 km northwest of Charleston, South Carolina. The initial shock had a duration of strong (felt) shaking of about 35 to 40 seconds (Ref. 18). A second strong shock occurred about eight minutes after the main shock. There were ten severe and numerous moderate aftershocks through September 30, 1886. Liquefaction appears to have been widespread in the meizoseismal area and was the cause of much damage. Building damage from ground shaking in Charleston was widespread, and 1400 chimneys were reported destroyed in the city (Ref. 18). Sixty people were killed in the earthquake. Most of the basic information on the effects of the earthquake appeared in an early report by Dutton (Ref. 34).

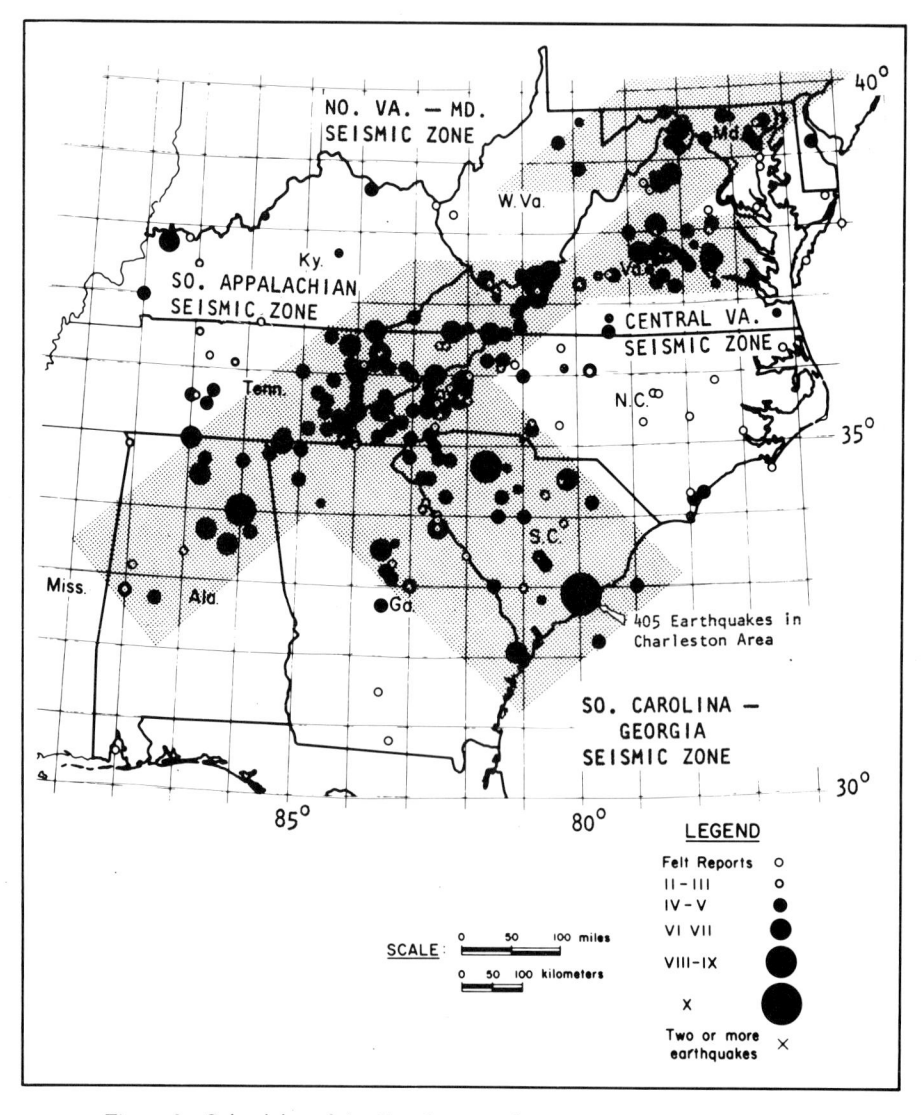

Figure 9a. Seismicity of the Southeast region, 1754-1970 (Refs. 32 and 33).

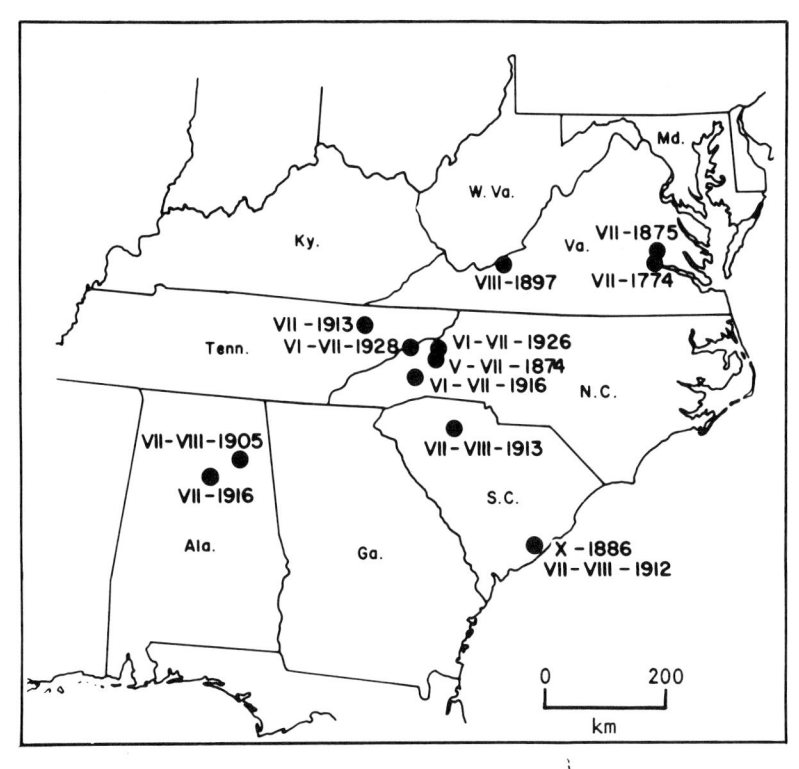

Figure 9b. Earthquakes in the Southeast region with Modified Mercalli intensities VII and larger (Ref. 32.)

There is evidence of foreshock activity a few days preceding the main shock, although it appears that the general level of seismicity in central and eastern South Carolina was very low during the 15 years before the 1886 shock (Ref. 35). For a period of ten years following 1886 there was also a very low level of seismicity in South Carolina and adjacent states except in the epicentral area of the 1886 shock. The epicentral area continues to be active at present but at a very low magnitude level (Ref. 36).

The effects of the 1886 earthquake have recently been restudied by Bollinger, who has assigned the main shock a maximum MMI of X and a magnitude (m_b) of 6.8, roughly equivalent to an M_s magnitude of 7.7 (Ref. 37). The intensity X was assigned principally on the basis of geological effects. Damages to

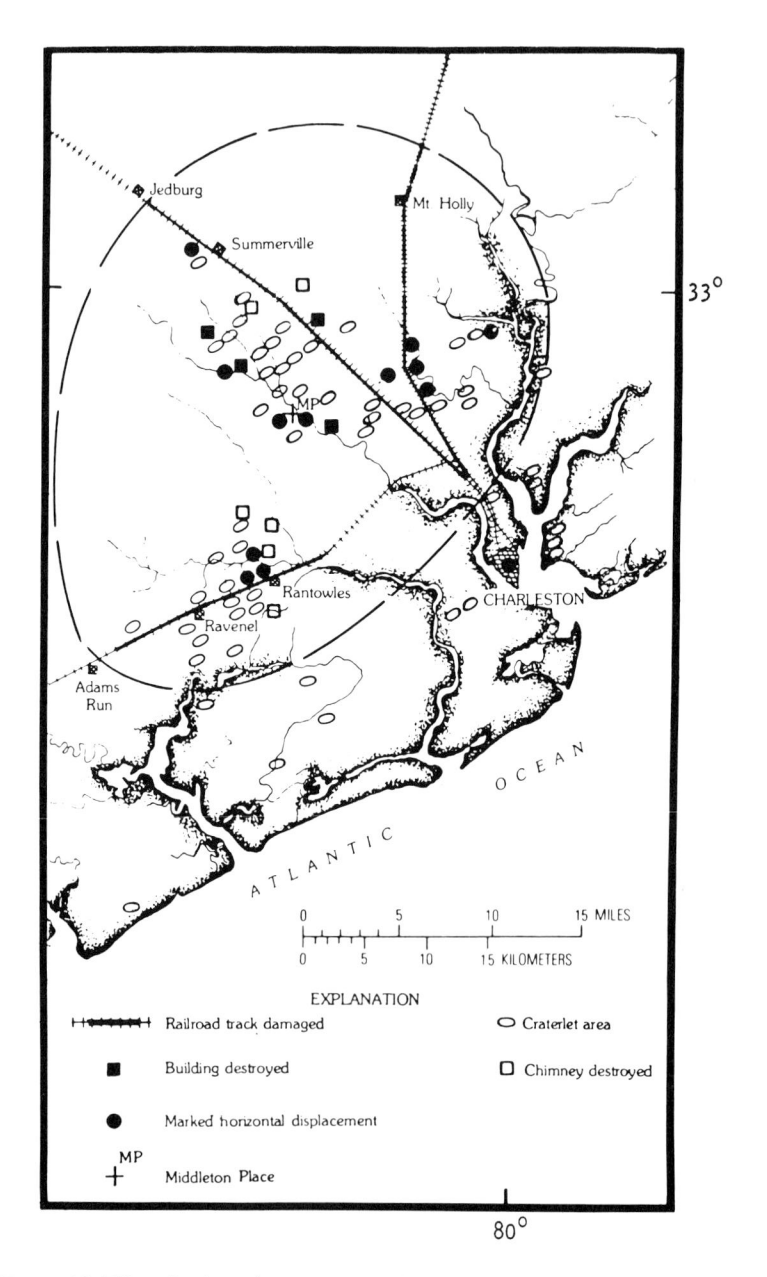

Figure 10. Effects in the epicentral area of the 1886 Charleston, South Carolina earthquake (Ref. 37).

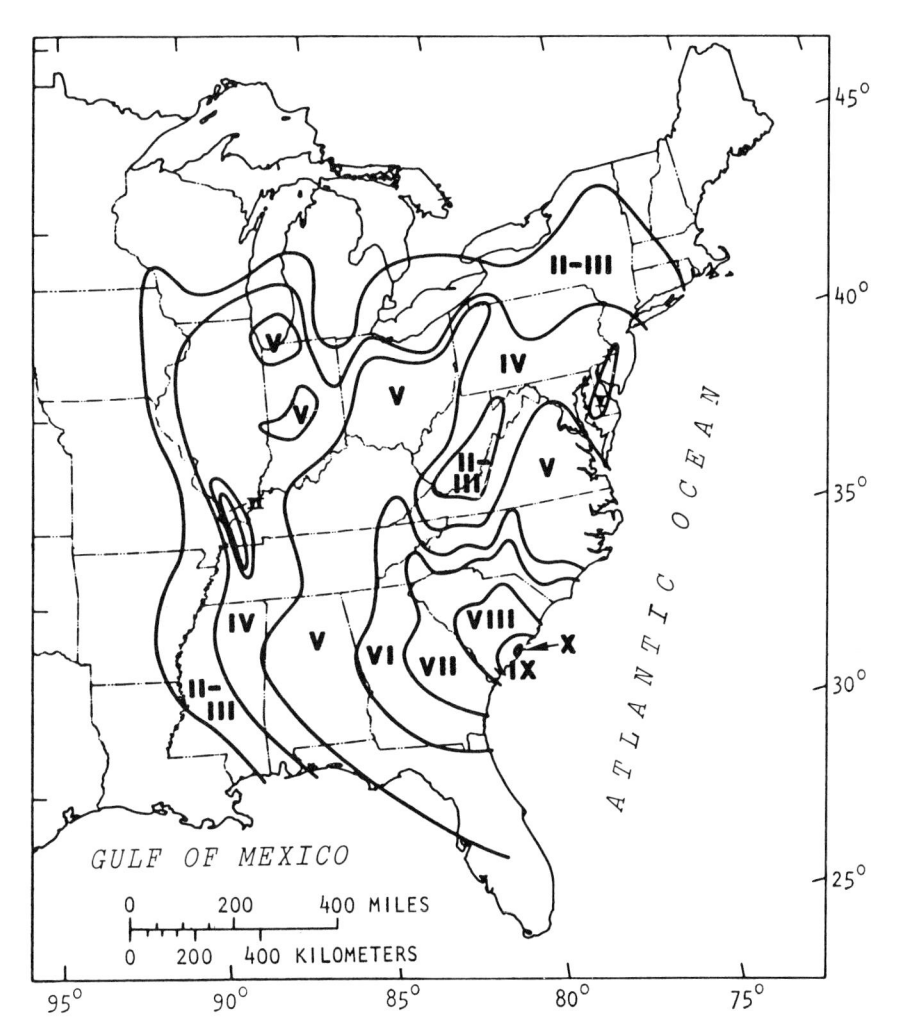

Figure 11. Isoseismal map of the 1886 Charleston, South Carolina earthquake (Ref. 37).

structures in Charleston were judged by Bollinger to be at the intensity IX level. Effects in the epicentral area are shown in Fig. 10; Fig. 11 is an isoseismal map of the earthquake. Felt areas of earthquakes in the southeastern United States have been estimated to be, on the average, about ten times larger than for earthquakes of comparable magnitude in California.

Table 3 gives the distribution of earthquakes by maximum MMI intensity in the Southeast, and Table 4 lists the shocks with maximum MMI's of VII and greater. Probably the second largest shock to occur in the region was the Giles County, Virginia earthquake of May 31, 1897, for which a magnitude (m_b) of 5.8 has been computed (Ref. 38). The shock was felt from Georgia to Pennsylvania and from the Atlantic Coast to Indiana and Kentucky. A recent hazard assessment was made of this earthquake by Bollinger (Ref. 39). A review of seismotectonic setting and seismicity by Bollinger and Wheeler concluded that the orientation of the zone appears to be related to features below the Appalachian overthrust belt (Ref. 40).

Table 3

Distribution of Earthquakes by Maximum Intensity in the Southeast Region*
(through 1976)

Maximum MM Intensity	Number
V	133
VI	70
VII	10
VIII	2
IX	0
X	1

*Source: Ref. 32 and U.S. Geological Survey earthquake data file. Earthquakes with epicenters believed to be offshore are not included.

A number of ideas have been advanced to explain the nature of earthquake occurrence in the southeastern United States. It has been noted (Ref. 41) that the South Carolina-Georgia seismic zone lies along an extension of the strike of the Blake Spur Fracture Zone (Fig. 8), although no evidence of this feature is known onshore.

It has also been suggested that the activity along the southern Atlantic coastal plain is related to reactivated old faults that were formed during the opening of the Atlantic Ocean in Mesozoic time. Thus, it may be possible that older fault zones control the

Table 4

*Important Earthquakes of
the Southeast Region*

Date (GMT)	Location	Maximum MM Intensity (I₀)*	Magnitude (Mₛ)	m_b
Feb 21, 1774	Eastern VA	VII		
Feb 10, 1874	McDowell County, NC	V-VII		
Dec 22, 1875	Arvonia, VA area	VII		
Aug 31, 1886	Near Charleston, SC	X	(7.7)	(6.8)
Oct 22, 1886	Near Charleston, SC	VII		
May 31, 1897	Giles County, VA	VIII	(6.3)	(5.8)
Jan 27, 1905	Gadsden, AL	VII-VIII		
Jun 12, 1912	Summerville, SC	VI-VII		
Jan 1, 1913	Union County, SC	VII-VIII	(5.7-6.3)	
Mar 28, 1913	Near Knoxville, TN	VII		
Feb 21, 1916	Near Asheville, NC	VI-VII		
Oct 18, 1916	Northeastern, AL	VII		
Jul 8, 1926	Mitchell County, NC	VI-VII		
Nov 2, 1928	Western NC	VI-VII		

*Source: Intensities are from Ref. 32. Bracketed magnitudes indicate they were estimated from intensity. The m_b magnitude for the 1886 earthquake is from Ref. 37 and for the 1897 earthquake from Ref. 38. The M_s magnitudes were estimated from intensity using $M_s = 1 + 2/3\ I_o$.

location of current seismicity. There is evidence that Mesozoic rifting occurred in the Charleston area as it did at other locations in the Atlantic coastal area (Ref. 42). Recently, northeast-trending, high angle reverse faults active in Cenozoic time have been identified on land near Charleston and offshore to the southeast. Other recent data from deep seismic reflection profiling suggest that the crystalline rocks of the southern Appalachians have been thrust at least 260 km to the west and that they overlie sedimentary rocks over an extensive area of the central and southern Appalachians (Ref. 43). It is not confirmed at present whether or not the overthrusting is present in the Charleston area. In the Giles County, Virginia area, monitoring of recent seismicity has shown that earthquakes occur as deep as

25 km. The proposed overthrusting in the Appalachians in general is believed to have a thickness from 6 to 15 km but only about 6 km in the Giles County locale, so that most of the earthquakes in the Giles County area are well beneath the thrusts.

This discussion raises an important question concerning the uniqueness of the tectonic setting of the 1886 Charleston earthquake, the largest historical earthquake in the area and the earthquake that dominates the seismicity of the southeastern coastal area. The suggestion has been made that earthquakes of the Charleston type could occur over a rather wide area of the Appalachians and along the east coast of the United States. However, no unifying theory of earthquake occurrence seems to be currently acceptable, and the geologic origins of earthquakes in this area remain an open question.

CENTRAL REGION

The seismicity of the central region is dominated by the large earthquakes that have occurred in the Mississippi River Valley in much the same way that the seismicity of the southeastern United States is dominated by the Charleston earthquake of 1886. Exclusive of Alaska, the earthquakes that occurred in the Mississippi Valley in 1811 and 1812 rank as the largest known shocks in North America since European settlement. The 1811-12 earthquake sequence has been extensively investigated by Nuttli (Refs. 44, 45) and the following discussion is taken principally from his comprehensive study.

The earthquake sequence began with a major shock on December 16, 1811 and was followed by numerous aftershocks (Table 5). An isoseismal map prepared by Nuttli is shown in Fig. 12. Soil liquefaction as well as regional subsidence and uplift was widespread. Local landsliding was common, especially along the rivers. A number of islands in the Mississippi River disappeared. The approximate distribution of sand blows (sand blows are the result of ejections of sand and water associated with soil liquefaction) is shown in Fig. 13. The other two principal shocks of the series occurred on January 23 and February 7, 1812.

Masonry and stone structures were damaged to distances of 250 km. Chimneys were destroyed in Louisville, Kentucky about 400 km from the earthquakes. Lesser chimney damage was found at distances of over 600 km. The earthquakes were felt south to the Gulf Coast, southeast to the Atlantic and northeast to Quebec. No reports of the earthquakes were available to the west, and this is reflected in Fig. 12. Nuttli (Ref. 46) places the epicenter of the December 16, 1811 shock near the southern edge of the area of soil liquefaction shown in Fig. 13. The third and largest shock of the series occurred on February 7, 1812, and was probably located near the north edge of the area of soil liquefaction, perhaps 10 to 20 km west of New Madrid. The

January 23 event is believed to have occurred roughly equidistant between the first and third shocks, but its epicenter is largely speculative.

The December 16, 1811 earthquake had (with the possible exception of the shock on February 7, 1812) the largest potential damage area and felt area known in the earthquake history of the United States. The area of potential damage (taken as the area shaken at an intensity level of VII or greater) has been estimated as 600,000 km^2 (Ref. 44). For comparison, a reasonable extrapolation of the intensity VII and greater area of the 1964 Alaska earthquake yields an area of about 210,000-250,000 km^2. The 1906 San Francisco earthquake had an area with intensity greater than or equal to VII of about 30,000 km^2.

In contrast to the usual occurrence of a single principal shock followed by a series of aftershocks, the 1811-12 earthquake series had three large shocks, each of which was followed by aftershocks, many of which were very large. Table 5 indicates that there were on the order of 15 aftershocks greater than M_s = 6.

Table 5

Principal Shocks and Aftershocks in the Mississippi Valley 1811-12 Series*

Principal Shocks	Magnitudes m_b	M_s	Maximum MMI
Dec 16, 1811	7.2	8.6	XI
Jan 23, 1812	7.1	8.4	X-XI
Feb 7, 1812	7.4	8.7	XI-XII

Aftershocks

— Five earthquakes with M_s magnitudes between 7 and 8
— Ten earthquakes with M_s magnitudes between 6 and 7
— 2000 earthquakes during the winter of 1811-12 that were strong enough to be felt in Louisville, Kentucky approximately 320 km from the meizoseismal area of the principal shocks

*Source: Ref. 45.

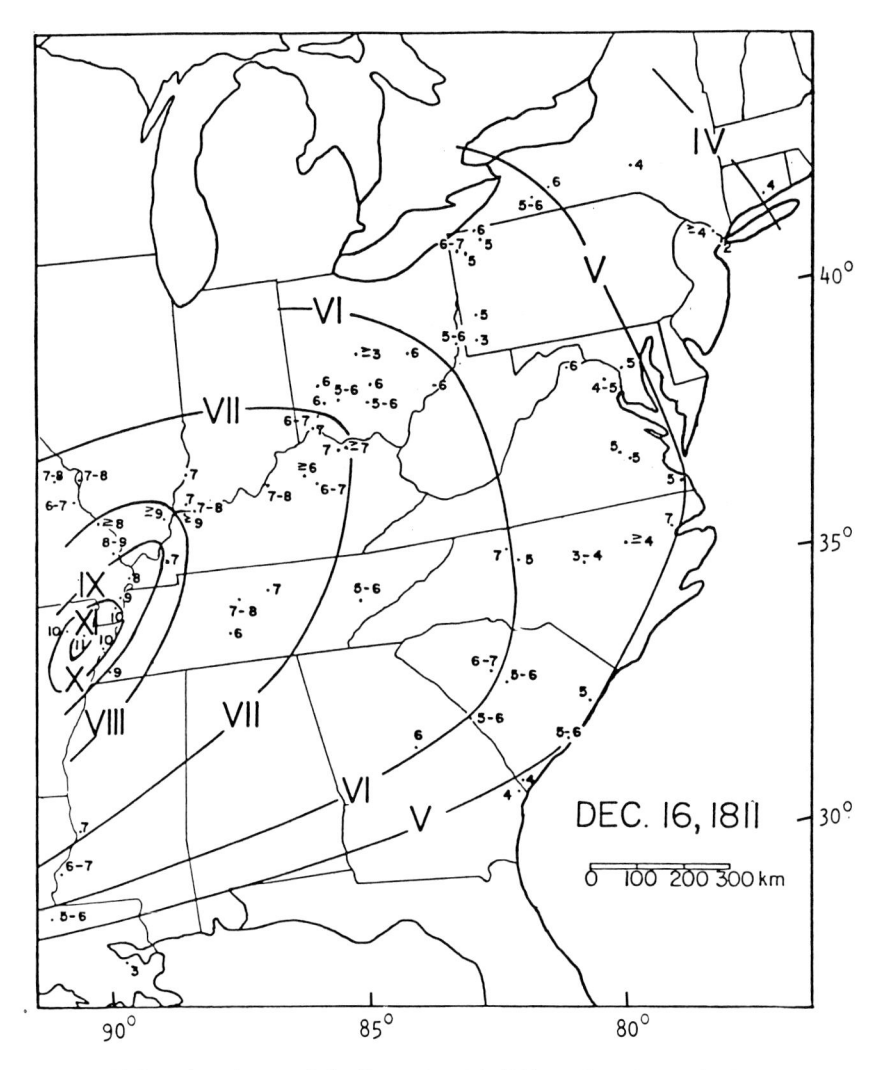

Figure 12. Isoseismal map of the December 16, 1811 earthquake (Ref. 45). The arabic numbers give the Modified Mercalli intensities at each data point.

There were more than 1600 aftershocks large enough to be felt in the first three months following the December 16, 1811 event. About as many earthquakes occurred in the Mississippi Valley area in these three months as occurred in southern California in the 40-year period from 1932 through 1972; aftershocks continued until at least 1817 (Ref. 47). The locations of these aftershocks are not known, but it is possible that they occurred over a considerably larger area than did the three main shocks.

Fuller (Ref. 47) has described the widespread uplift and subsidence in the New Madrid area, although the association of some of the features described by Fuller with the 1811-12 series has been questioned. No surface faults clearly associated with the earthquake activity in the New Madrid seismic zone or the

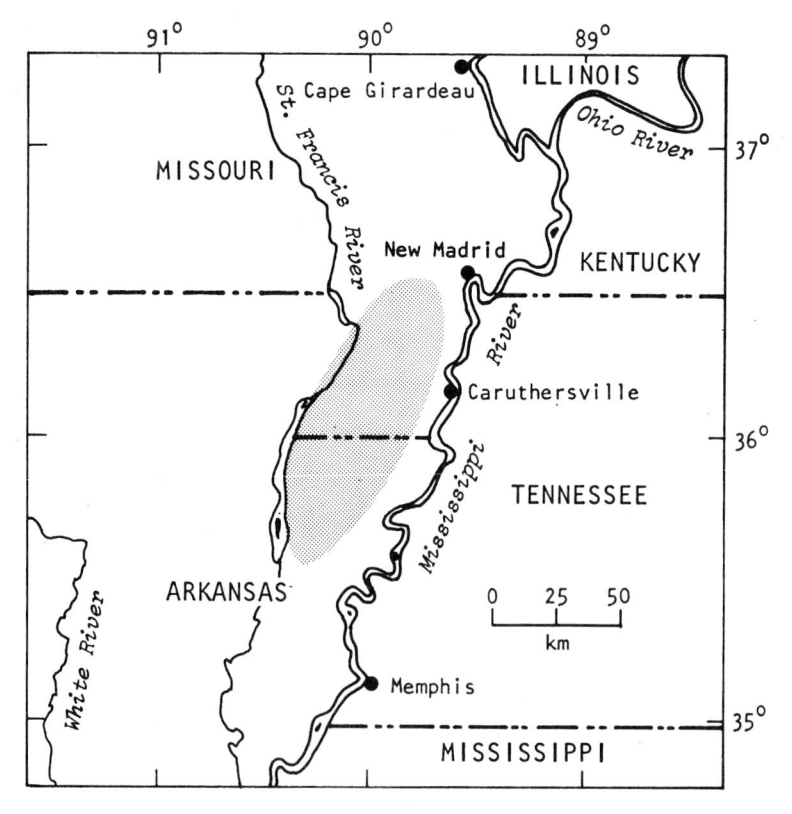

Figure 13. Epicentral region of the 1811-12 Mississippi Valley earthquakes. The shaded area contains numerous sand blows, clear evidence of soil liquefaction resulting from the earthquakes. The December 16, 1811 initial large event is believed to have occurred near the southern end of the zone of liquefaction; the final large earthquake (February 7, 1812) possibly 10 km west of New Madrid (Ref. 46).

surrounding area have been identified. This suggests that at least the large earthquakes are not extremely shallow (< 15 km) and probably occurred at depths of 15-30 km. Depths greater than about 30 km for large shocks would not seem likely, based on the observation that the larger earthquakes known to have occurred in the Midwest (in 1811-12) had long aftershock sequences, a characteristic of large shallow earthquakes. Further evidence that these were shallow earthquakes is that many of the small earthquakes of recent years have been well located and are known to be quite shallow.

Important earthquakes in the central region exclusive of the 1811-12 series, are listed in Table 6. The distribution of earthquakes by maximum MM intensity is given in Table 7, including the three large shocks of the 1811-12 series but excluding the aftershocks. The completeness of the historical record of seismicity for the central region is related to the westward settlement of the area. The seismicity is reasonably well known for nearly 200 years in the eastern portion and for only about 100 years in the area west of Missouri, about 95° west longitude. It is believed that earthquakes in the region with epicentral Modified Mercalli intensities of VI or greater have been completely reported after settlement of these areas (Ref. 48). However, the instrumental network of seismograph stations over much of the central region still remains inadequate to provide significant seismotectonic data, outline active faults and improve the evaluation of the seismic hazard.

Figure 14 shows the seismicity of the central region through 1976. One significant earthquake occurred slightly north of the area shown in Fig. 14 on May 15, 1909 at approximately 49°N, 104°W on the United States-Canadian border. The maximum intensity was V-VI, and a magnitude (m_b) of 5.5 has been estimated from intensity data. This earthquake and the seismicity of northeastern Montana and Saskatchewan have been discussed by Horner and Hasegawa (Ref. 49).

The majority of the epicenters are based on analysis of earthquake reports, that is, intensity data. A limited network of seismograph stations in the Mississippi Valley was established by St. Louis University in the 1930's, but it was not until the

Table 6

*Other Important Earthquakes of the
Central Region Through 1980**

Date (GMT)	Location	Maximum MM Intensity (I_o)	Magnitude (m_b)
Jun 9, 1838	Southern IL	VIII	(5.7)
Jan 5, 1843	Near Memphis, TN	VIII	(6.0)
Apr 24, 1867	Near Manhattan, KS	VII	(5.3)
Oct 22, 1882	West Texas	VII-VIII	(5.5)
Oct 31, 1895	Near Charleston, MO	VIII-IX	(6.2)
Jan 8, 1906	Near Manhattan, KS	VII-VIII	(5.5)
Mar 9, 1937	Near Anna, OH	VIII	5.3
Nov 9, 1968	Southern IL	VII	5.5
Jul 27, 1980	Near Sharpsburg, KY	VII	5.1

*Source: Ref. 50, except for the 1980 event taken from Ref. 13 and the 1882 event from Ref. 179. Brackets indicate magnitude was estimated from intensity. The 1811-12 series of earthquakes is not included in this list (See Table 5).

Table 7

*Distribution of Earthquakes by Maximum
Intensity in the Central Region**
(through 1976)

Maximum MM Intensity	Number
V	275
VI	114
VII	32
VIII	5
IX	1
X	0
XI	2
XII	1

*Source: Ref. 50 with minor changes and additions.

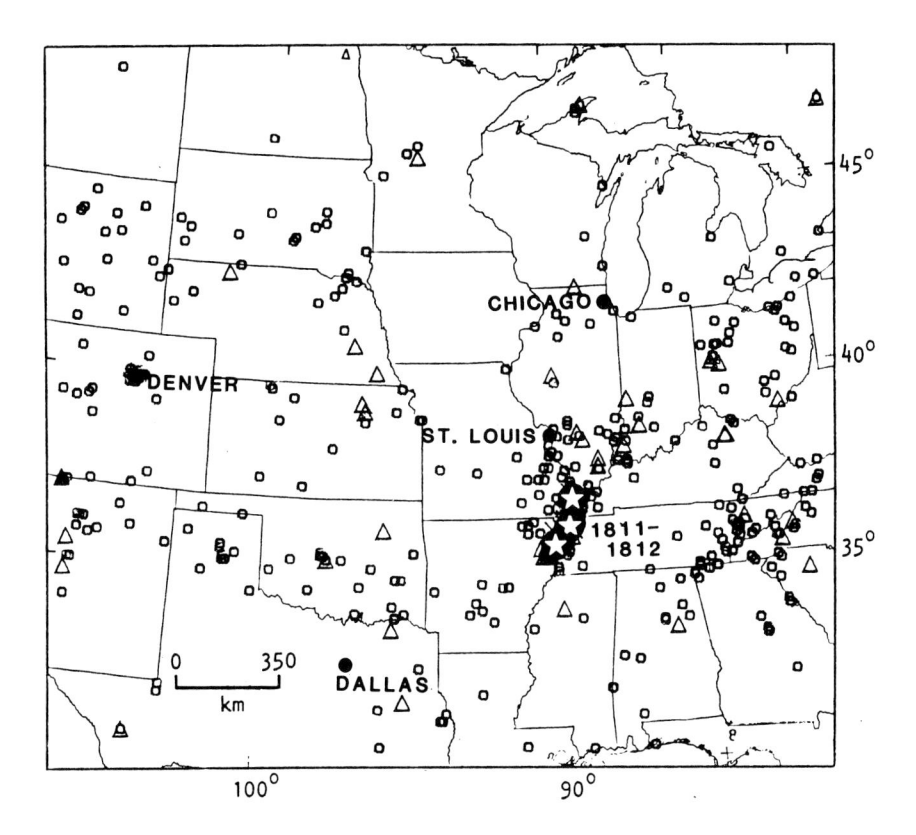

Figure 14. Seismicity of the Central region, 1811-1976. The data are taken principally from Ref. 46 with minor changes and additions. The stars represent earthquakes with maximum MM intensities of IX or greater; triangles represent earthquakes with maximum intensities of VII-VIII; squares represent earthquakes with maximum intensities of V-VI.

installation of a modern seismograph network by St. Louis University in cooperation with the U.S. Geological Survey in 1974 that the spatial distribution of seismicity was clarified in the New Madrid zone. Microearthquakes — that is, earthquakes with magnitudes less than about 3 — located during a 21-month period of operation of this network are shown in Fig. 15. An inspection of Fig. 15 shows that prior to the analysis of data from the microearthquake network, the seismicity of the New Madrid Zone revealed only a diffuse zone of activity.

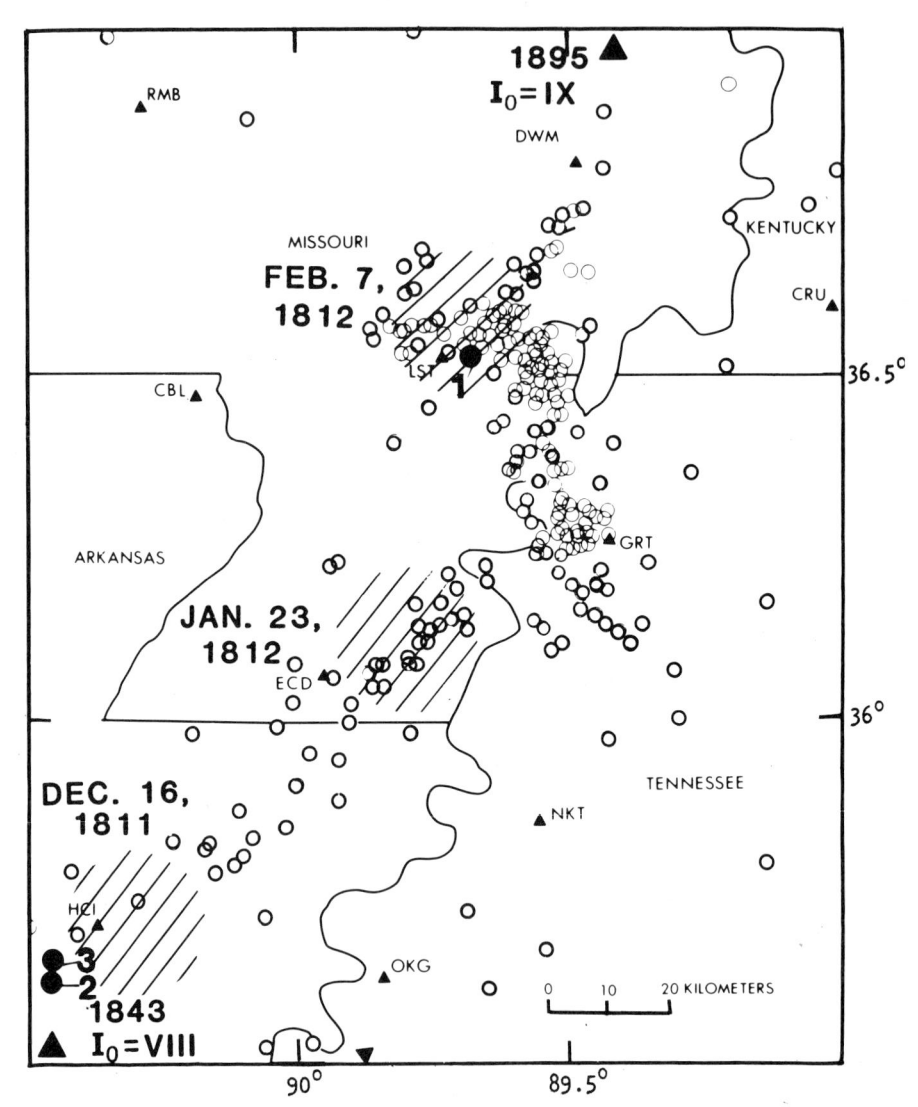

Figure 15. Epicenters located by the St. Louis University microearthquake network during a 21-month period July 1, 1974 to March 31, 1976. Larger shocks during the period are numbered and shown as solid circles: (1) June 13, 1975, m_b = 4.3; (2) March 25, 1976, m_b = 5.0; (3) March 25, 1976, m_b = 4.5. The approximate epicentral areas of the three principal shocks of 1811-12 are shown as hachured areas. Two other significant earthquakes in the zone that occurred in 1843 and 1895 are shown as triangles. The microearthquake data are taken from Ref. 51.

The pattern of contemporary seismicity emerges clearly even though less than two years of data from the microearthquake network are shown. Figure 15 also shows the approximate epicenters of the three main shocks of the 1811-12 series and two other important earthquakes that occurred in the zone, one in 1843 (I_o = VIII) and the other in 1895 (I_o = IX). It is interesting to note that the epicenters of the 1843 and the 1895 earthquakes are located respectively near the south and north ends of the principal microearthquake activity. Figure 16 shows the relationship of the microearthquake data to regional geological features and illustrates many of the results of increased research in the past ten years in the Mississippi River Valley. The microearthquakes define line segments: (1) from Marked Tree, Arkansas to Ridgely, Tennessee, striking northeast; (2) from south of Ridgely to about 20 km west of New Madrid, Missouri, striking slightly west of north; and (3) from the vicinity of New Madrid to Charleston, Missouri, striking northeast (Ref. 42).

The principal seismicity occurs in the northern Mississippi embayment, a south plunging wedge of Cenozoic sedimentary rocks. Specifically, the New Madrid seismic zone is located in a southwest-northeast striking graben or downthrown block about 70 km wide and at least 200 km long. This graben is believed to have formed in the late Precambrian era during a period of continental rifting (Refs. 53, 54, 55). The rift structure was defined principally on the basis of the interpretation of magnetic data (Refs. 53, 56). Magnetic data have also identified several plutons (intrusions of igneous rock into the basement) that seem to bound the rift, particularly on the north, west and east sides. Seismic reflection data indicate repeated faulting along preexisting zones of weakness in the rift (Ref. 57). Very old Continental rifts appear to be important in concentrating seismicity not only in the New Madrid area but elsewhere in the central and eastern United States, because they are weak zones in the crust that may be reactivated repeatedly if they are properly oriented in the present stress field. For a more complete discussion, see Refs. 42, 58, and 59.

An unanswered question is whether the rift structure identified in the Mississippi Valley is unique in the central United States or

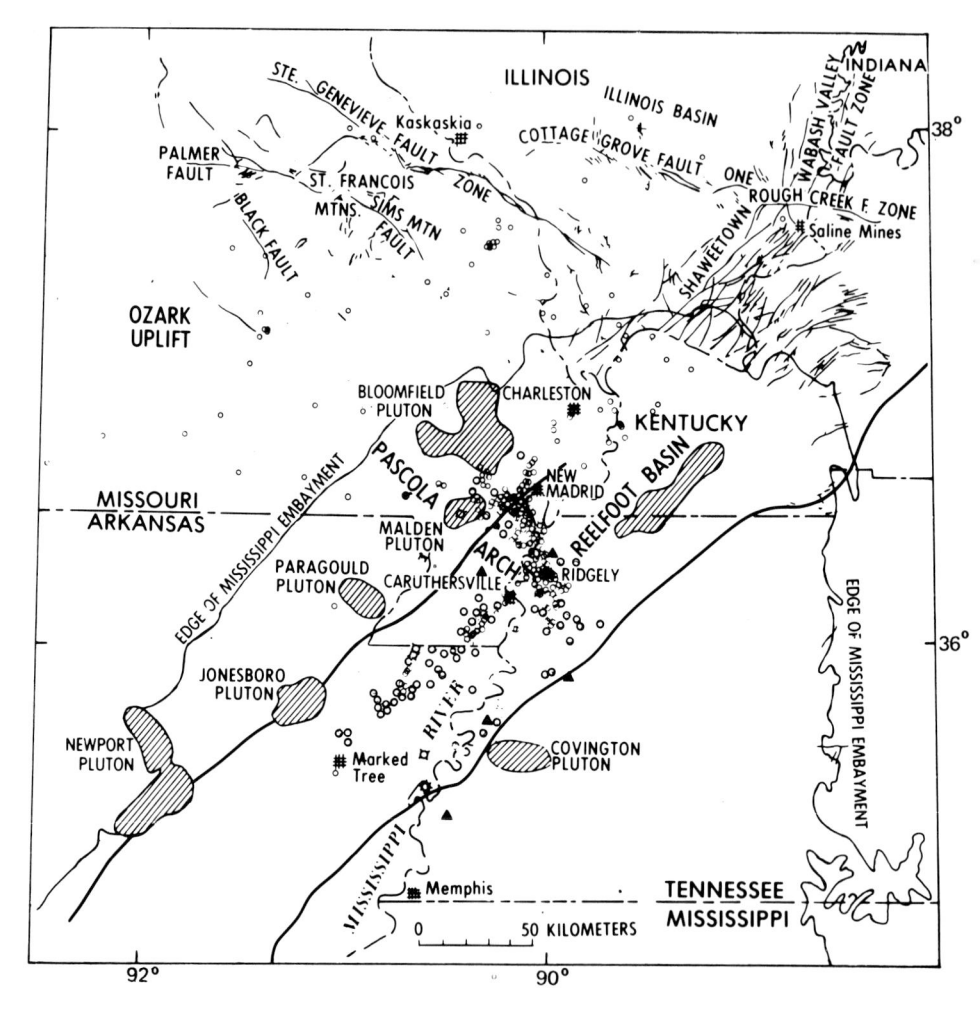

Figure 16. The Mississippi Valley seismic region (Ref. 42) showing microearthquake activity (circles, from Ref. 52) and faults (Ref. 56). Graben boundaries are shown by heavy lines, plutons by hachured areas (Ref. 52).

whether other similiar structures may exist. Damaging earthquakes have also occurred sporadically in the central United States in the Wabash Valley area northeast of the New Madrid zone, along the Ouachita-Wichita mountains in Oklahoma, in northeast Kansas and southern Nebraska, in northern Illinois, in a relatively small area in western Ohio near

Anna and in northern Kentucky. Conclusive correlations between the occurrence of earthquakes in these areas and geologic structures have not been demonstrated, although a number of seismotectonic relationships have been postulated. In contrast to Fig. 14., which shows that earthquakes with epicentral intensities of V and VI are quite widely distributed throughout the central United States, only earthquakes with maximum MM intensities of VII or greater are plotted in Fig. 17. Note that damaging shocks outside the Mississippi Valley are uncommon in the Midwest.

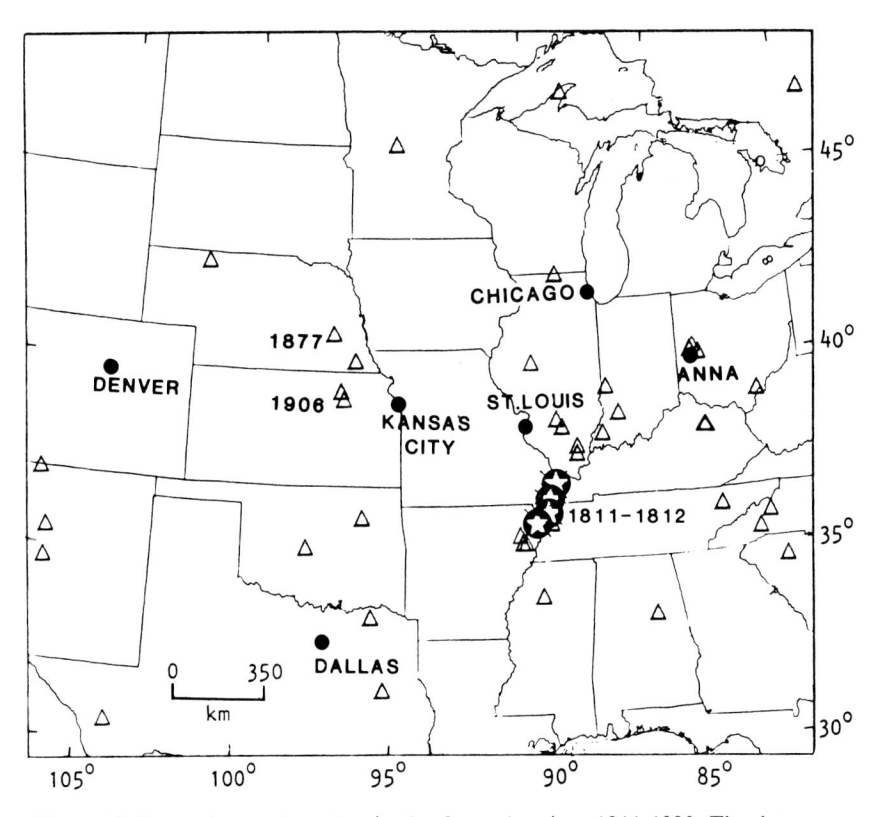

Figure 17. Damaging earthquakes in the Central region, 1811-1980. The data are taken principally from Ref. 46 with minor changes. The triangles represent earthquakes with maximum MM intensities of VII and VIII. The four large stars represent the three main shocks of the 1811-12 sequence and the maximum intensity IX earthquake of 1895 near Charleston, Missouri.

The concentration of seismicity in western Ohio near Anna has been studied by Bradley and Bennett (Ref. 61). Damage occurred in Anna and in some of the surrounding towns in 1931, and again six years later as a result of earthquakes on March 2 and March 8, 1937. The 1937 shocks caused extensive chimney damage in Anna and the surrounding area. Plaster was cracked in Fort Wayne, Indiana and motion was felt in tall buildings in Chicago, Milwaukee, and Toronto (Ref. 18). The area has been active at least since 1875, when a damaging shock occurred, and it remains active today. The Anna, Ohio earthquakes are in the vicinity of the intersection of the Cincinnati, Kankakee and Finley arches which began to develop by differential subsidence in Late Cambrian-Early Ordovician time (Ref. 62). There is also minor activity along the east and west flanks of the Finley arch in northern Ohio.

Damaging earthquakes have occurred north of the New Madrid area in the Wabash Valley in southeastern Illinois and southwestern Indiana. The largest event (m_b = 5.5), on November 9, 1968, was felt over 23 states and Canada. Damage was minor, probably because depth of focus was about 20 km (Ref. 63). Minor damage was reported in Evansville, Illinois, St. Louis, and Chicago. It was felt in tall buildings in Mobile, Alabama, southern Ontario, and Boston (Ref. 48). Speculation exists regarding the relationship between the Wabash Valley seismicity and the more intense activity in the New Madrid area to the south. Nuttli (Ref. 48) believes that earthquake magnitudes should not exceed m_b = 6.5 in the Wabash Valley on the basis of the shorter length of the known faults in the Wabash Valley, compared to the longer inferred fault lengths (about 200 km) in the New Madrid zone.

The Wichita-Ouachita structural system extends westward across central Oklahoma and has contained the loci of a number of small earthquakes and a damaging shock near El Reno, Oklahoma in 1952 (m_b = 5.5; I_o = VII). There were chimney damage, cracked walls and glass breakage in El Reno, and minor damage in Oklahoma City and Ponca City (Ref. 18). Beneath the Mississippi Embayment, the southwest extension of the rift structure in the New Madrid zone may be intersected or terminated in southeastern Arkansas by the buried Wichita-Ouachita structural system. The role that this intersection plays in controlling seismicity is at present not well understood. Figure 14 shows a clustering of Intensity V and VI earthquakes in this vicinity.

The significant earthquake activity in Kansas has been located along or near the Nemaha ridge, a buried basement feature in east central Kansas that strikes approximately N10°E of north and extends into Oklahoma where it intersects the Wichita-Ouachita buried mountain front. Activity is minor and not continuous along the feature. The seismic history of the state has been outlined by Merriam (Ref. 64) and recently reviewed and updated by Dubois and Wilson (Ref. 65). The two largest earthquakes, both with maximum MM intensities of VII-VIII, occurred on April 24, 1867 and on January 8, 1906 in the vicinity of Manhattan, Kansas. The locations are not very well known and there is no consensus regarding which shock is the larger. Dubois and Wilson (Ref. 65) consider the 1867 earthquake to have a maximum MM intensity of VII-VIII, while Nuttli and Herrmann (Ref. 50) assign this earthquake a maximum intensity of VII. The reverse interpretation applies to the 1906 earthquake (Ref. 65, intensity VII; Ref. 50, intensity VII-VIII). The larger shocks are believed to be associated with the Humboldt fault zone, along the eastern edge of the Nemaha Ridge (Ref. 66).

The Sharpsburg, Kentucky earthquake (m_b = 5.2, M_s = 4.7) of July 27, 1980 is the third largest earthquake in the eastern half of North America since 1961. The focal depth was about 12-15 km and the faulting has been interpreted as a northeast right lateral strike slip (Refs. 67, 68, 69). The shock was felt over about 600,000 km^2 and was assigned a maximum intensity of VII. Losses in excess of $3 million (1980) were estimated. The greatest damage was at Maysville, Ohio, but chimney damage was widespread. The earthquake occurred in an area of previously low seismicity (Ref. 70).

In summary, several generalizations are possible concerning the seismicity of the central region: (1) earthquakes of small magnitude are scattered throughout the region and may be associated with local distortions of the generally compressive

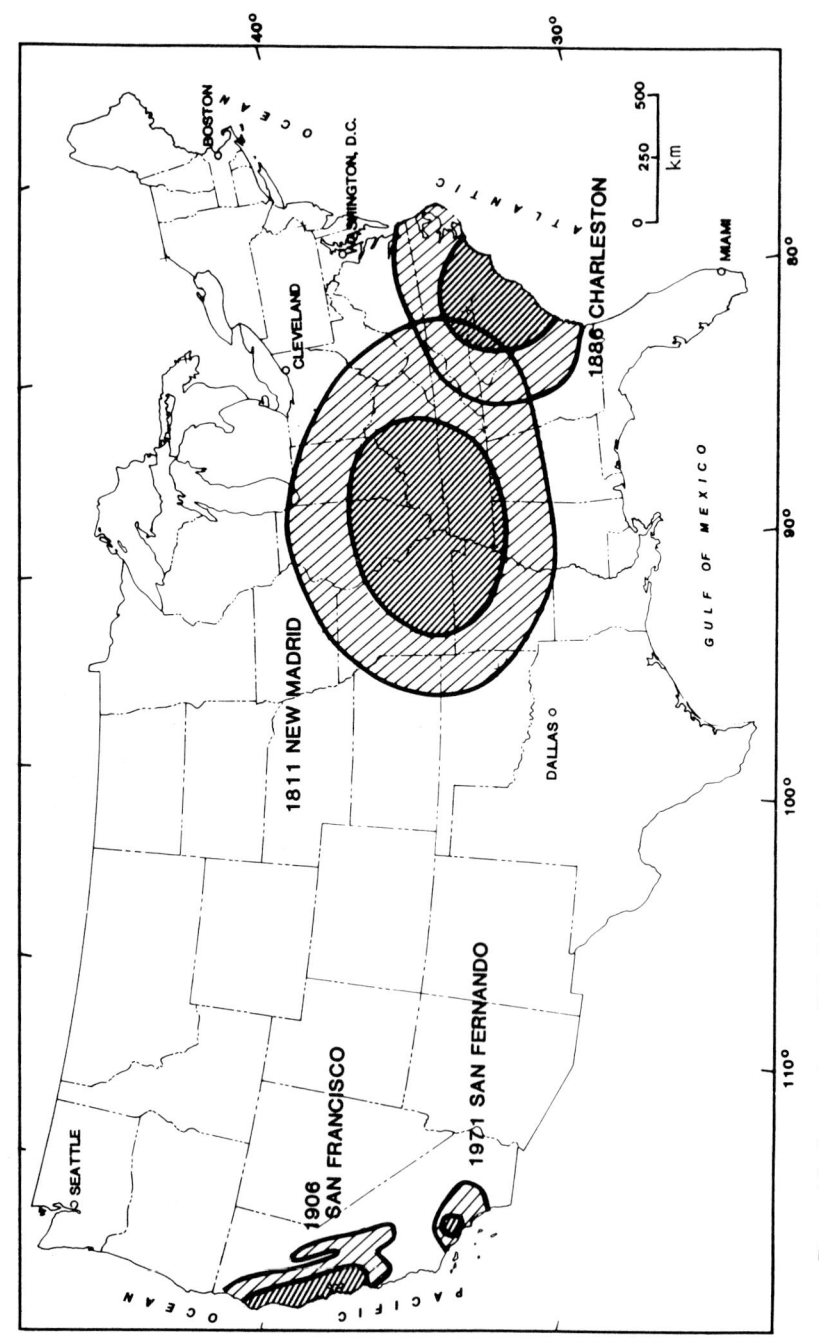

Figure 18. Comparison of Modified Mercalli intensity data for four earthquakes. The dark areas are intensity VIII and greater; the light areas are intensity VI-VII. The magnitudes of the earthquakes are: 1971 (M_s = 6.3); 1906 (M_s = 8.3); 1886 (M_s = 7.8); 1811 (M_s = 8.6). Adapted from Ref. 48.

stress field known to exist throughout the Midwest (Ref. 58); these earthquakes would be expected to be less than $m_b = 5.0$; (2) with the exception of the rift structure identified in the New Madrid area, other significant seismicity appears to be located on the flanks or crests of basement highs; (3) the major seismicity is associated with the rift structure identified in the New Madrid area. The boundaries of the rift are at present somewhat uncertain. The structure may extend further to the south than presently recognized and may extend into the St. Genevieve fault system to the northwest and possibly into the Wabash Valley area; (4) focal mechanisms of earthquakes derived in recent years from seismograms of small earthquakes are consistent with the regional compressive stress in the midwest United States; (5) the attenuation of seismic waves in the Midwest is clearly anomalously low with regard to attenuation in the western United States. This low attenuation also occurs in the eastern United States but perhaps to a lesser extent. Figure 18 shows the differences in attenuation between eastern-central U.S. earthquakes and western earthquakes; (6) upper bound magnitudes for earthquakes in the central region (and the east) are difficult to establish, although it is reasonable to assume that the 1811-12 earthquakes in the Mississippi Valley are near or at the upper bound for that area; and (7) the average recurrence rates for large shocks are not very well known because of the small data sample. Estimates based on the magnitude distribution of historical earthquakes indicate average recurrence intervals of 500-1000 years for the Mississippi Valley. This does not imply that, at present, the time to the next large shocks in the Mississippi Valley is necessarily 300 to 800 years since it is known that the interoccurrence times of large shocks vary over wide limits. The 500-1000 year estimate is only a rough estimate of the *average* recurrence interval for large shocks in the Mississippi Valley.

WESTERN MOUNTAIN REGION

Important earthquake activity in the Western Mountain region has occurred in the Yellowstone Park-Hebgen Lake area, in western Montana, in the vicinity of the Utah-Idaho border and sporadically along the Wasatch fault (Fig. 19 and Table 8). The Oct. 1980 issue of the *Bulletin of the Seismological Society of America* is devoted to special papers on seismicity of the Wasatch Fault and Great Basin-Sierra Nevada boundary.

The largest earthquake in historic times was the $M_s = 7.1$ shock of August 17, 1959, which occurred in the Yellowstone Park-Hebgen Lake area. The earthquake caused a massive landslide blocking the Madison River about 10 km downstream from Hebgen Dam. The impounded water reached depths of 53 meters. Hebgen Dam was damaged and there was also extensive road damage throughout the area. Twenty-eight lives were lost, principally in landslides. The main shock produced an east-west striking normal fault with 6.7 meters of displacement (Ref. 73). A long aftershock sequence followed the main shock. The aftershock zone is elongated in an east-west direction along the strike of the faulting. Focal depths of aftershocks ranged from about 24 km to very shallow shocks. In general, the aftershocks were deeper in the western portion of the aftershock zone.

Damaging earthquakes occurred in the Helena, Montana area in 1925 and again in 1935 when two shocks with magnitudes greater than 6 occurred within 13 days. The 1925 earthquake probably occurred in the Clarkson Valley about 80 km southeast of Helena and was slightly larger than the 1935 shocks. The series of earthquakes in 1935 was somewhat closer to Helena than those in 1925 (Ref. 74). Important early strong ground motion data were obtained at Helena from some of the later shocks of the 1935 series. None of the earthquakes produced surface faulting. The area has been relatively quiet since 1935. Figure 19 shows that there is a diffuse zone of seismicity from the Yellowstone Park-Hebgen Lake area through the vicinity of Helena, Montana and northwest to the

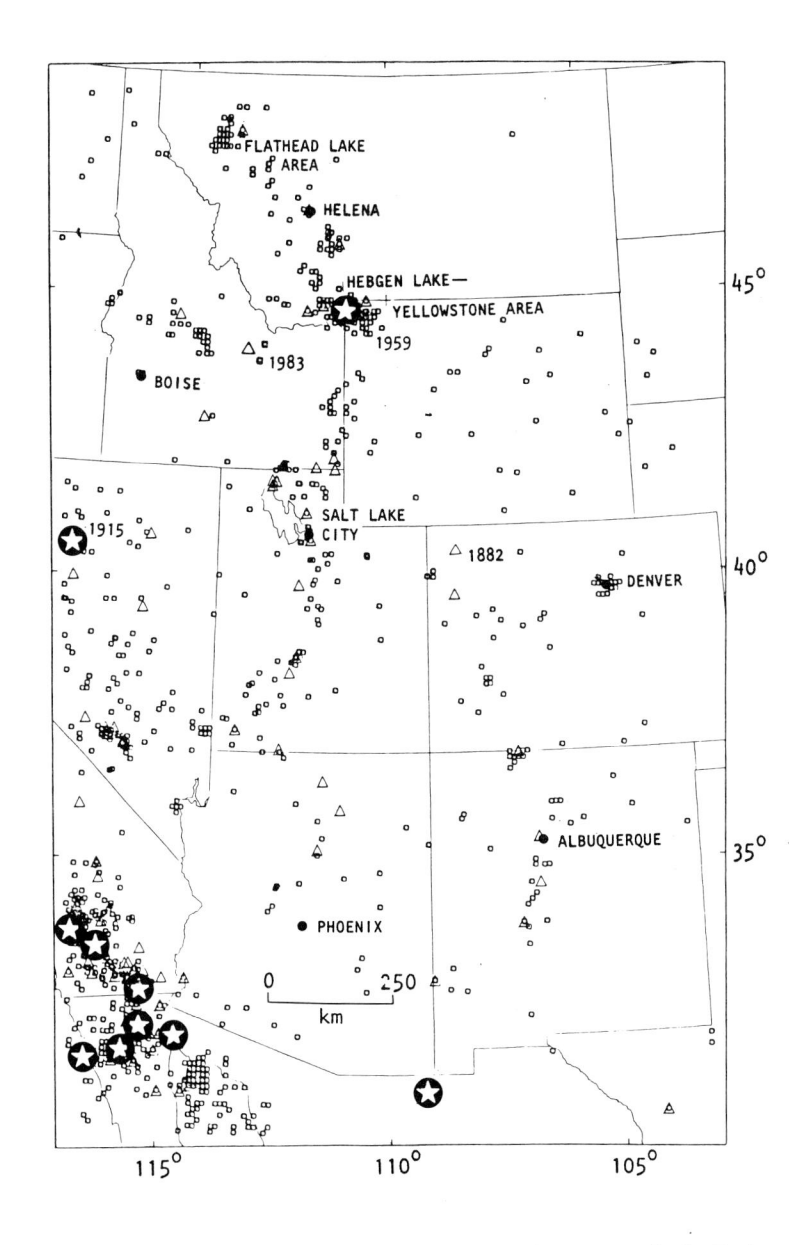

Figure 19. Seismicity of the Western Mountain region (data principally from Ref. 71). Stars represent earthquakes with maximum intensities of IX or greater; triangles represent earthquakes with maximum intensities of VII-VIII; and squares represent earthquakes with maximum intensities of V-VI.

Table 8

Important Earthquakes of the Western Mountain Region*

Date (GMT)	Location	Maximum MM Intensity (I_o)	Magnitude M_s**
Nov 9, 1852	Near Ft. Yuma, AZ?	VIII?	
Nov 10, 1884	Utah-Idaho border	VIII	
Nov 14, 1901	About 50 km east of Milford, UT	VIII	
Nov 17, 1902	Pine Valley, UT	VIII	
Jul 16, 1906	Socorro, NM	VIII	
Sep 24, 1910	Northeast Arizona	VIII	
Aug 18, 1912	Near Williams, AZ	VIII	
Sep 29, 1921	Elsinore, UT	VIII	
Sep 30, 1921	Elsinore, UT	VIII	
Oct 1, 1921	Elsinore, UT	VIII	
Jun 28, 1925	Near Helena, MT	VIII	6.7
Mar 12, 1934	Hansel Valley, UT	VIII	6.6
Mar 12, 1934	Hansel Valley, UT	VIII	6.0
Oct 19, 1935	Near Helena, MT	VIII	6.2
Oct 31, 1935 (Aftershock)	Near Helena, MT	VIII	6.0
Nov 23, 1947	Southwest MT	VIII	
Aug 18, 1959	West Yellowstone-Hebgen Lake	X	7.1
Aug 18, 1959 (Aftershock)	West Yellowstone-Hebgen Lake	VI	6.5
Aug 18, 1959 (Aftershock)	West Yellowstone-Hebgen Lake	VI	6.0
Aug 18, 1959 (Aftershock)	West Yellowstone-Hebgen Lake	VI	6.0
Aug 18, 1959 (Aftershock)	West Yellowstone-Hebgen Lake	VI	6.5
Mar 28, 1975	Pocatello Valley, ID	VIII	6.1M_L
Jun 30, 1975	Yellowstone National Park	VIII	6.4M_L
Oct 28, 1983	Lost River Mtns., ID	VII est.	7.3

*Source: U.S. Geological Survey Data File and Ref. 71.
**M_s magnitudes are from Ref. 18. M_L magnitudes are from Ref. 71; for an explanation of the M_L scale used in Utah, see Ref. 72.

Flathead Lake area. The Flathead Lake area has displayed increased seismic activity since 1964. The geological sources of this activity have not been identified.

The seismicity in the Great Basin is concentrated along the east and west margins of the Basin — on the east along the Wasatch fault zone and in the Nevada seismic zone on the west side of the Great Basin. The only historical earthquake to produce surface faulting in the eastern Great Basin is the shock of March 12, 1934 (M_s = 6.6) in Hansel Valley near the north shore of the Great Salt Lake. There was up to 50 cm of vertical displacement on normal faults and considerable liquefaction and subsidence (Ref. 73). A second shock (M_s = 6.0) occurred in the same area about three hours after the initial earthquake. Damage was relatively low because the area was lightly populated. An M_s = 6.0 magnitude earthquake with a depth of about 9 km occurred north of Hansel Valley in Pocatello Valley, Idaho in 1975. A foreshock of M_L = 4.2 preceded the mainshock by 22 hours. The shock produced no surface faulting, but a north-trending fissure zone having no apparent displacement was found in the south central part of Pocatello Valley. More than 600 aftershocks were located using a network of portable seismographs (Ref. 75). The mechanism of the main shock was normal faulting on a north trending plane. Figure 20 shows the historical seismicity of the area and the relationship of the 1975 earthquake to the earlier activity in Hansel Valley. The nearest population center to the 1975 main shock is Malad City, Idaho, where about 40 percent of the chimneys on old buildings were damaged and about 300 homes had some degree of damage (Ref. 18).

The strongest known earthquake in the Western Mountain region in 24 years occurred on October 28, 1983 at the base of Borah Peak in the Lost River Mountains in Idaho. The event (M_s = 7.3) was felt in nine states, and within two hours, more than 15 aftershocks were recorded. Heavy damage occurred — mostly to older buildings — in Challis, 30 miles away from the epicenter, and in Mackay, 20 miles away. Damage was estimated at $2.5 million (Ref. 12). Two schoolchildren were killed by a falling storefront in Challis, and became the first such deaths in the United States since the 1971 San Fernando Valley earthquake.

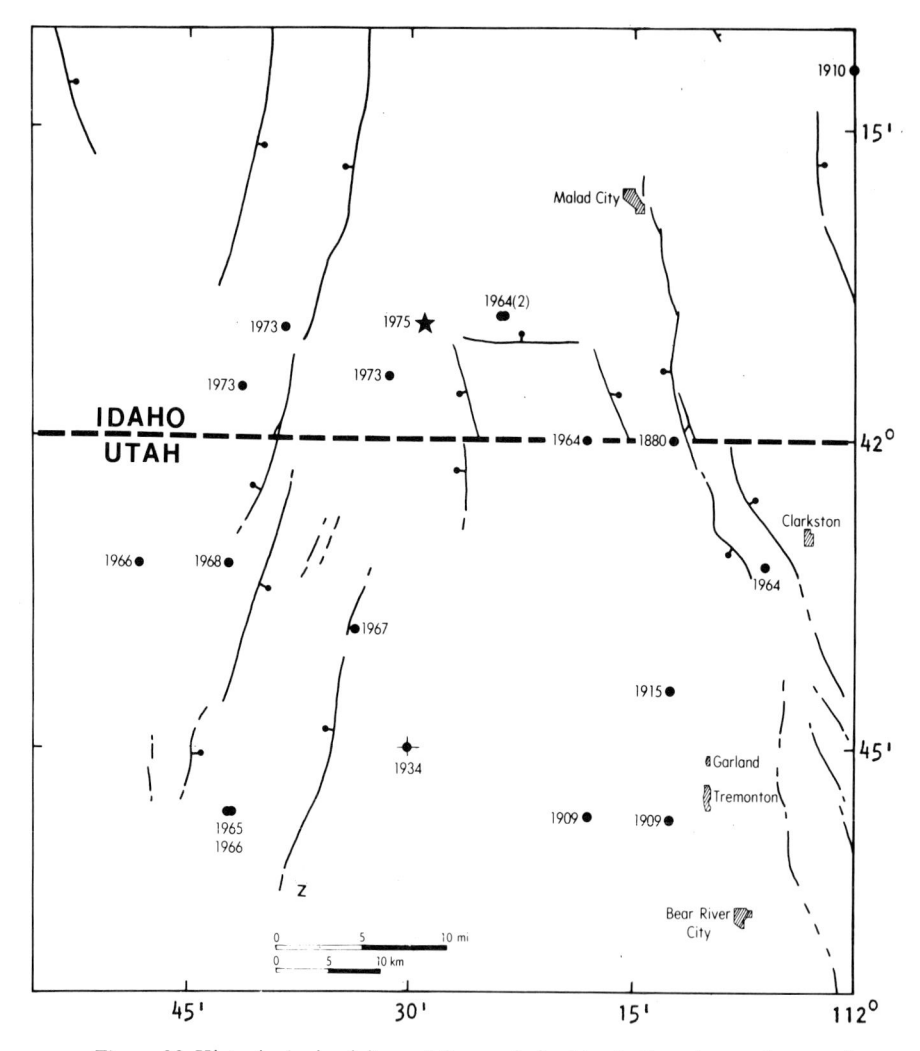

Figure 20. Historical seismicity and Cenozoic faulting in the epicentral areas of the 1975 Idaho (M_s = 6.1) and 1934 Utah (M_s = 6.6) earthquakes (Ref. 76). The epicenters of the 1934 and earlier earthquakes are only approximate.

Surface faulting of 22 miles has been identified. Normal faulting predominated, with a maximum vertical displacement of 2.5 m; however, left lateral fault movement was also noted at a number of locations along the fault. The average vertical displacement on the fault was about 1.5 m.

Only a few earthquakes producing significant damage have occurred along the Wasatch fault system in north central Utah. However, a M_s = 5.7 shock on the Utah-Idaho border in 1962 did cause damage along the east side of Cache Valley from Logan to Lewiston, Utah. The most severe damage was in Richmond, Utah where nine houses were declared unsafe and chimney loss was more than 50 percent. Four schools in Cache County were seriously damaged. This 1962 earthquake was the most damaging shock in Utah since the 1934 Hansel Valley earthquake. While there is sporadic seismic activity along the Wasatch fault in central Utah, most of the shocks have been small and have produced little damage. Figure 21 displays the space-time distribution of earthquakes within a 10 km epicentral distance of the Wasatch fault from the Utah-Idaho border to Levan, Utah, a distance of about 270 km, for the period July, 1962 through June, 1978. Note that gaps in seismic activity occur along the fault and that some sections of the fault are active, then become quiescent while other sections become active (Ref. 77). Large earthquakes (M_s = 6.5 to 7.5) have occurred on the Wasatch fault in Holocene time (past 10,000 years), but not since 1850 when historical seismicity has been documented in Utah (Ref. 78). There are several possible explanations for the lack of large earthquakes on the Wasatch fault in historic time: (1) strain energy is accumulating and a large shock can be expected in the future; (2) some other mechanism is relieving the strain energy, for example, seismic creep on the fault or crustal rebound; (3) the seismic activity is migrating eastward; or (4) the Wasatch fault is at present less active than in the past, and less strain energy is accumulating.

Significant earthquakes did occur in south central Utah in the Sevier Valley near Elsinore in 1921 when three damaging shocks with magnitudes near 6 occurred within about 25 hours. Aftershocks continued for four months. There was widespread damage at Elsinore, Richfield and Monroe, Utah.

In southern Utah the seismicity splits into two trends, with one trend swinging to the southwest toward southern Nevada and the other trend continuing southward into Arizona. Major seismic activity has not occurred in Arizona historically but in 1887, a

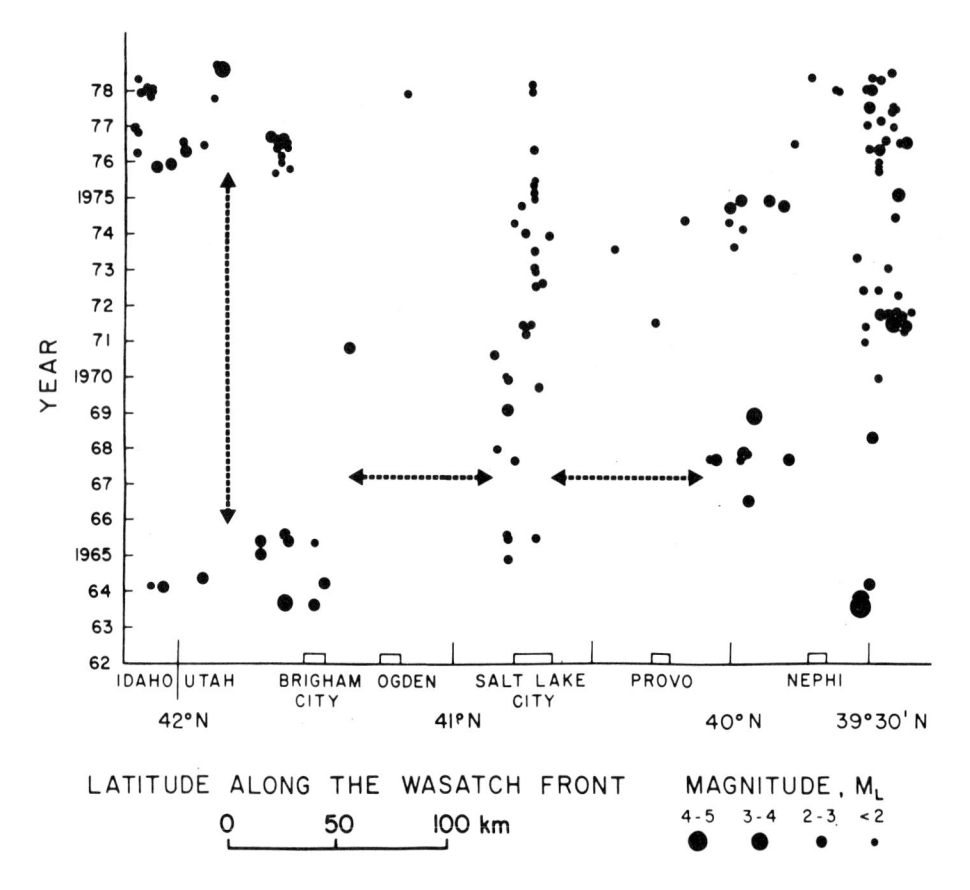

Figure 21. Space-time distribution of earthquakes within a 10 km epicentral distance of the Wasatch fault from July 1962 through June 1978. Time and space gaps in the seismicity are shown by arrows (Ref. 77).

large earthquake (I_o = XI-XII) occurred in Sonora, Mexico in the San Bernardino Valley along the western front of the Sierra Madre Mountains. Faulting associated with the earthquake extends southward from about 8 km south of Douglas, Arizona for a distance of 50 km to Batepito, Sonora (Ref. 79). The earthquake has important implications for seismic risk evaluations in southern Arizona. Most of the seismicity in Arizona occurs in a broad northwest trending area of late Cenozoic faulting in a transition zone between the Colorado

Plateau, with a crust about 40 km in thickness, and the Basin and Range Province, characterized by a 30 km or less crustal thickness and high heat flow. The relationship of this seismic activity to the Sonora earthquake of 1887 is not clear. Recently, the historical record of earthquakes in Arizona has been reviewed and a number of small to moderate magnitude additional earthquakes have been added to the catalog (Ref. 80).

Microearthquake investigations in 1978 and 1979 in the Chino Valley, about 20 km north of Prescott, Arizona indicated that the level of seismicity in the Chino Valley area was an order of magnitude less than in the Intermountain Seismic Belt in Utah. The m_b = 4.9 earthquake that occurred in the Chino Valley in 1976 was the largest earthquake in Arizona in the twenty-year period 1960-1980 (Ref. 81).

Seventy-six percent of the earthquakes in New Mexico confirmed for the time period 1849-1975 occurred in the Rio Grande rift, sometimes called the Rio Grande depression, graben or trench (Ref. 82). In New Mexico, the rift extends from the Colorado-New Mexico border to the Mexican border. It varies in width from about 15 to 100 km and is roughly centered on the Rio Grande River. The rift is characterized by grabens and extensional faulting. In the Socorro area and elsewhere there is evidence of rapid Holocene uplift and magma within the crust (Ref. 83). The highest level of historical seismicity present within the rift occurs between Albuquerque and Socorro (Ref. 84). An excellent summary of earthquake activity in New Mexico through 1977 is given in Ref. 85.

In northern New Mexico, a significant area of low to moderate seismicity occurs about 30 km west of the rift, well inside the boundary of the Colorado Plateau, near Dulce, New Mexico. The sequence of shocks beginning on January 23, 1966 with a mainshock of m_b = 5.5 caused about $200,000 of damage, principally at Dulce. The mainshock has been relocated at 36.98° N, 107.02° W at a depth of 3 km. No surface faulting was found (Ref. 86). It is interesting to note that although the earthquake occurred well inside the Colorado Plateau, the focal mechanism determined from instrumental data is consistent with east-west tensional forces found in the Rio Grande rift.

Elsewhere in the Western Mountain region, seismicity is sporadic and generally of low level. The concentration of earthquakes northeast of Denver, which began in 1962, is associated principally with the injection of liquids into a deep disposal well on the grounds of the Rocky Mountain Arsenal, northeast of Denver. The largest event occurred on August 9, 1967 (I_o = VI-VII) and was the largest earthquake in the Denver area in historic times (Ref. 87).

Figure 22 shows the earthquakes with maximum intensities of VII and greater, and Table 9 lists the distribution of earthquakes in the Western Mountain region by maximum intensity. Note that the largest historical earthquakes have occurred in the Yellowstone Park-Hebgen Lake area and in northern Sonora, Mexico. Since Holocene and Late Quaternary fault scarps are fairly widely distributed throughout the Western Mountain region, indicating occurrences of large earthquakes in the past 10,000-250,000 years, a reasonable conclusion is that the potential for large earthquakes may exist in a number of areas that have not as yet experienced them historically and that the recurrence rates of these large shocks throughout most of the area are low. The historical seismicity of northern Sonora, except for the 1887 earthquake, is not very well known. Better seismic monitoring of northern Sonora and southern Arizona is needed to evaluate the seismic hazard of the area.

Table 9

Distribution of Earthquakes by Maximum Intensity in the Western Mountain Region*
(through 1976)

Maximum MM Intensity	Number
V	474
VI	149
VII	26
VIII	22
IX	0
X	1

*Source: Ref. 71 with minor changes

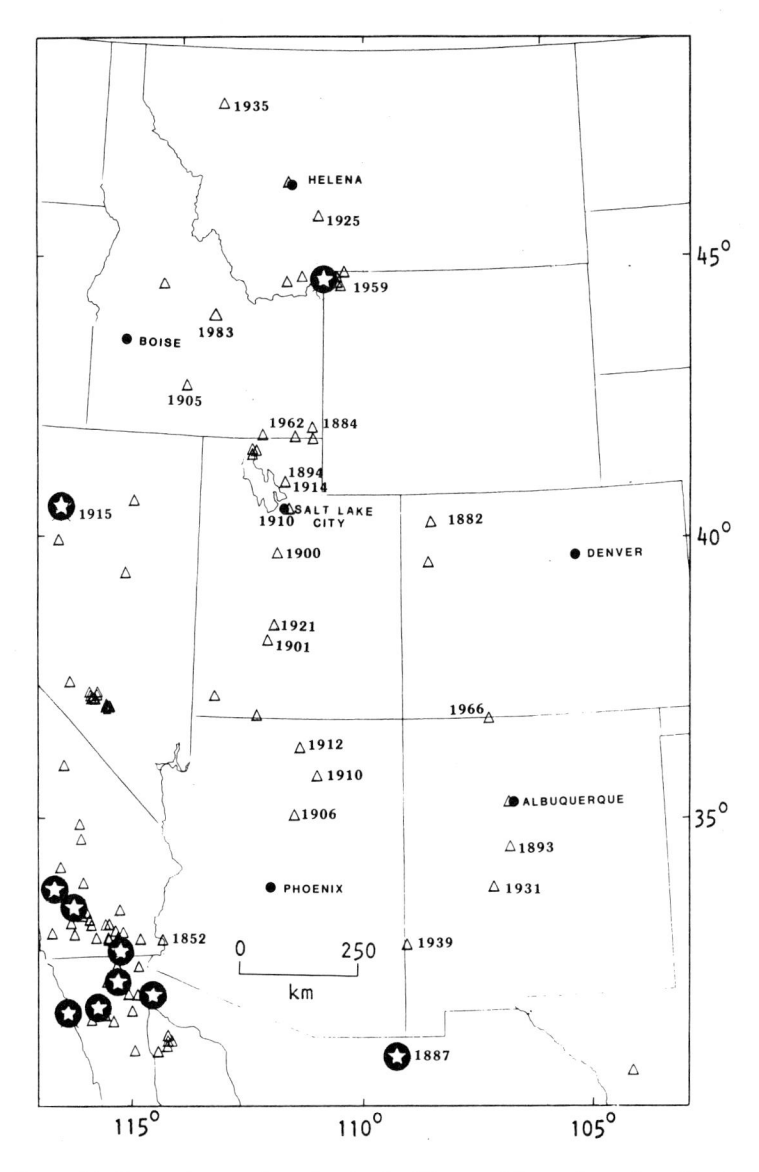

Figure 22. Damaging earthquakes in the Western Mountain region. The data are taken principally from Ref. 71. The triangles represent earthquakes with maximum MM intensities of VII and VIII. The star represents an earthquake with a maximum intensity of IX or greater.

The Wasatch fault represents an anomalous situation in that the recurrence rate for large shocks may be as high as one per 100-200 years on the fault, yet historically (130 years) none have occurred. Additional geological field work may reveal additional anomalous areas of this type.

PACIFIC REGION

The Pacific Region is subdivided here into two regions: California and western Nevada, and Washington and Oregon.

California and Western Nevada

The highest rates of seismic energy release in the United States exclusive of Alaska, occur in California and western Nevada. The entire area consists of a transition zone (which may extend east beyond California and Nevada to the Wasatch fault in central Utah) between the North American plate and the Pacific plate. One view of the tectonics is that western North America is within a broad belt of right lateral movement (west side moving north) which is associated with the differential movement between the North American and the Pacific plates. This type of movement has produced great strike-slip faults such as the San Andreas and related systems and also produces tensional features such as block faulting that occur in the Basin and Range province. Thus, a number of different styles of faulting which occur in the Basin and Range province* and west to the California coast can be explained by this general concept. Other questions such as the high heat flow in the Great Basin and the considerable width of the transition zone assumed in this hypothesis between the North American and Pacific plates are not so easily explained by such a simple concept. A careful review of current ideas has been presented by Smith and Eaton (Ref. 88).

Figure 23 shows the earthquakes with maximum MM intensity of VI or greater in California and western Nevada. For comparison, Fig. 24 shows only shocks of maximum intensity

---

The Basin and Range Province is used here to mean a physiographic province in the southwest United States characterized by a series of tilted fault blocks forming longitudinal asymmetric ridges or mountains and broad intervening basins (Ref. 95). A further description and map may be found in Ref. 96. The Great Basin is a subdivision of the Basin and Range Province.

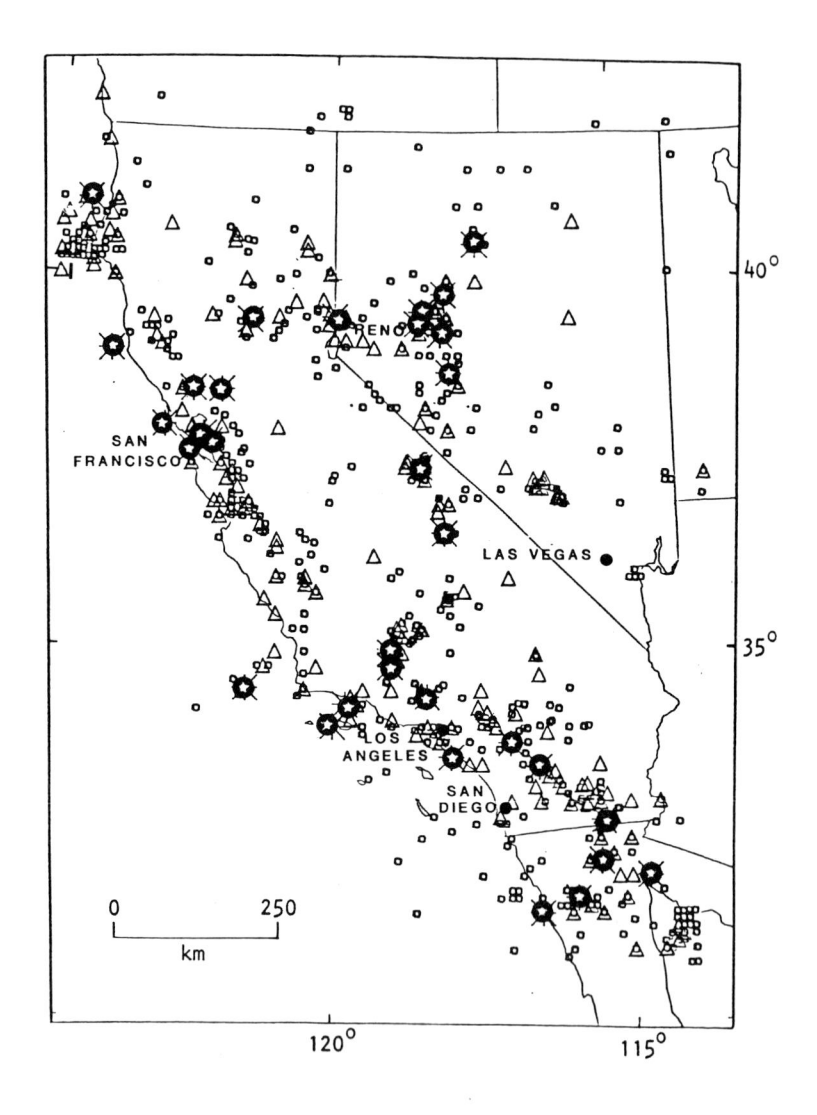

Figure 23. Seismicity of western Nevada and California, 1811-1976. The data are principally from Refs. 12, 89, 90 and 91 with minor changes and additions. Stars represent earthquakes with Modified Mercalli intensities of IX or greater, triangles represent shocks with maximum intensities of VII-VIII, and small squares represent shocks with maximum intensities of VI.

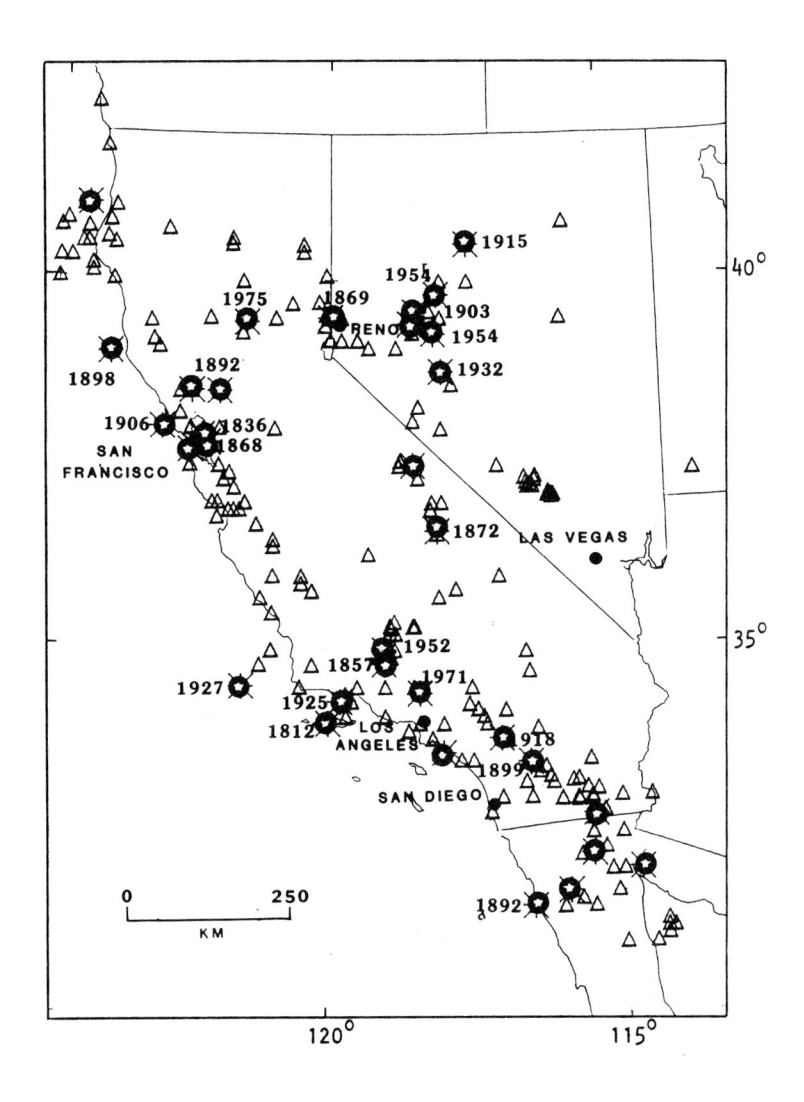

Figure 24. Earthquakes producing significant damage in California and western Nevada, 1811-1976. The data are principally from Refs. 12, 89, 90 and 91 with minor changes and additions. Stars represent earthquakes with Modified Mercalli intensities of IX or greater; triangles represent earthquakes with maximum intensities of VII-VIII.

VII and greater, with the most important earthquakes identified by year. Table 10 lists the major earthquakes, and the distribution of earthquakes by maximum intensity is given in Table 11.

Table 10

*Major Earthquakes of California and Western Nevada**

Date (GMT)	Location	Intensity (MM)	M_s
Dec 21, 1812	Santa Barbara Channel	(X)	
Jun 10, 1836	Hayward Fault, east of San Francisco Bay	IX-X	
Jun, 1838	San Andreas fault	X	
Jan 9, 1857	San Andreas fault, near Fort Tejon	X-XI	
Oct 21, 1868	Hayward Fault, east of San Francisco Bay	IX-X	
Mar 26, 1872	Owens Valley	X-XI	
Apr 19, 1892	Vacaville, CA	IX	
Apr 15, 1898	Mendocino County, CA	VIII-IX	
Dec 25, 1899	San Jacinto, CA	IX	
Apr 18, 1906	San Francisco	XI	8.3
Oct 3, 1915	Pleasant Valley, NV	X	7.7
Apr 21, 1918	Riverside County, CA	IX	6.8
Mar 10, 1922	Cholame Valley, CA	IX	6.5
Jan 22, 1923	Off Cape Mendocino, CA	(IX)	7.3
Jun 29, 1925	Santa Barbara Channel	VIII-IX	6.5
Nov 4, 1927	West of Point Arguello, CA	IX-X	7.3
Dec 21, 1932	Cedar Mountain, NV	X	7.3
Mar 11, 1933	Long Beach, CA	IX	6.3
May 19, 1940	Southeast of El Centro, CA	X	7.1
Jul 21, 1952	Kern County, CA	XI	7.7
Jul 6, 1954	East of Fallon, NV	IX	6.6
Aug 24, 1954	East of Fallon, NV	IX	6.8
Dec 16, 1954	Dixie Valley, NV (2 shocks)	X	7.3
Feb 9, 1971	San Fernando, CA	XI	6.4
Oct 15, 1979	Imperial Valley, CA	IX	6.6
May 2, 1983	Coalinga, CA	VIII	6.5

*Ref. 18 with minor changes. Parentheses mean the earthquake epicenter is offshore and the intensity is estimated from the epicentral area.

Most of the major earthquakes of California and western Nevada have produced surface faulting, but in general the faulting in the Basin and Range province appears to be more complex than that in western California. Most of the large earthquakes of coastal California have produced strike-slip faulting although often with a component of normal or reverse motion. The earthquakes in historic times that have produced surface faulting are shown in Fig. 25.

A good general description of the types of faulting found in California, their geomorphic expression and various features found in fault zones, is given by Richter (Ref. 2). The large earthquakes of 1903, 1915, 1932, 1934 and 1954 in the Nevada Seismic Zone in west central Nevada all produced fault scarps, but the faulting is often found to be complex. For example, the faulting associated with the 1915 Pleasant Valley, Nevada earthquake consists of four normal faults with mainly dip-slip displacement although right-slip components were present in some of the faults. However, the echelon pattern of rupture of the faults suggests deep left-slip deformation on the northeast trending zone. The sequence of large earthquakes that occurred in the Nevada seismic zone in 1954 produced normal and oblique

Table 11

*Distribution of Earthquakes by Maximum Intensity in California and Western Nevada**

Maximum MM Intensity	Number
V	1263
VI	487
VII	170
VIII	40
VIII-IX	2
IX	8
IX-X	3
X	5
X-XI	2

*Source: Ref. 18; only earthquakes with epicenters onshore are considered.

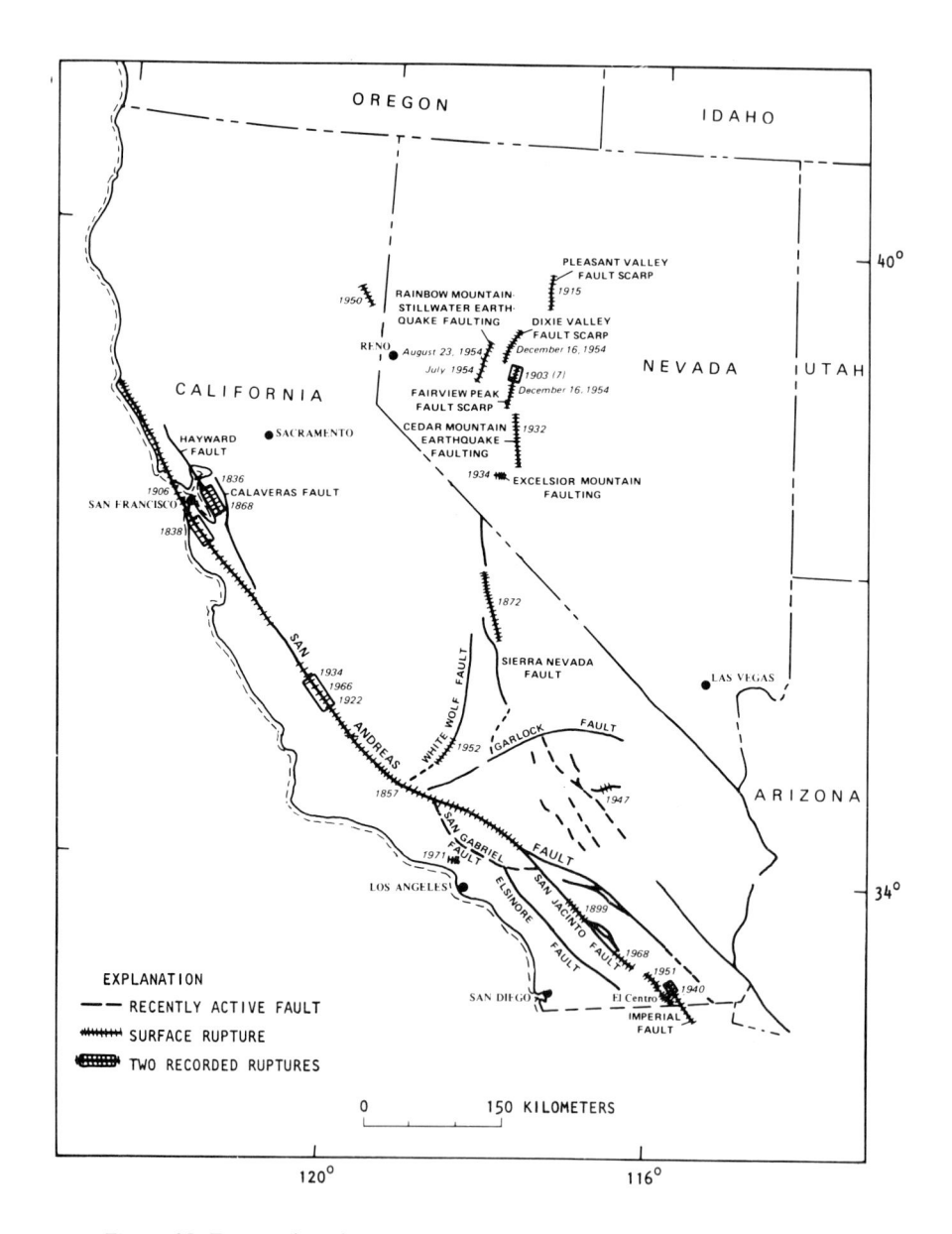

Figure 25. Faults with historic displacements in California and Nevada. The year of occurrence for selected large earthquakes is shown.

faulting over a region of deformation 93 km in length and 47 km in width (Ref. 92).

One of the largest earthquakes in California occurred near Lone Pine in the Owens Valley in 1872. Faulting took place along the Sierra Nevada fault zone for a distance of about 150 km in a nearly north-south direction. While Owens Valley is bounded on the east and west by steeply dipping normal faults of great displacement, movements during the 1872 earthquake were predominantly strike-slip with a small component of normal faulting. The motion was probably right lateral (Ref. 93). Although this earthquake occurred in a sparsely populated area, the percentage of casualties and damage to the existing structures was high. Richter lists 27 deaths in a population of 250-300 at Lone Pine (Ref. 2). Of 59 houses (for the most part of adobe construction), 52 were destroyed.

The area north of the 1872 Owens Valley earthquake is structurally complex, with earthquake swarms south and east of Mono Lake which may be related to volcanic processes. The area west of Walker Lake is an area of low seismicity where deformation appears to be taking place primarily by folding rather than faulting. Considerable activity has occurred in a zone from about 20 km southeast of Lake Tahoe to Reno, Nevada along a fault zone that marks the eastern boundary of the Sierra Nevadas. A good review of this seismicity is given by Van Wormer and Ryall (Ref. 94).

An important difference in the temporal and spatial occurrence of earthquakes in the Basin and Range Province and California has been recognized for some time. In the Basin and Range Province and particularly in the Nevada Seismic Zone in central and western Nevada, the recurrence interval for large earthquakes on any particular fault segment may be quite long, of the order of thousands of years (Refs. 98, 99). On the other hand, for California's San Andreas fault system, the recurrence rate for large shocks is of the order of 100 years (Ref. 100). The large area of Holocene faulting in central and western Nevada indicates that major earthquakes have occurred over a wide area during the past 10,000 years but that the interoccurrence times of activity on single fault segments are long.

The coastal areas of California are part of the active boundary between the Pacific and North American plates. Earthquake activity extends from the Gulf of California-Salton trough area northward along the San Andreas fault system to Cape Mendocino. The seismicity is not confined to a single fault but is distributed over a fault system 50 to 100 km wide that accommodates most of the motion between the plates (Figs. 23 and 24). Major earthquakes have occurred on the main trace of the San Andreas fault but important damaging shocks have also occurred on relatively obscure faults (obscure at least, until the earthquake occurred). Examples are the 1971 San Fernando earthquake and the 1952 series of shocks on the White Wolf fault in Kern County, California. Repeated damaging shocks, most recently in 1979, have occurred on the Imperial and related faults in the Imperial Valley. Wallace (Refs. 99, 100, 101) has described the behavior of different segments of the San Andreas fault (Fig. 26).

During the 1906 San Francisco earthquake, a segment of the San Andreas fault from Cape Mendocino to Los Gatos ruptured, with right lateral displacements of up to 7 meters. An earlier large earthquake had occurred on this segment in 1838. Since 1906 this portion of the San Andreas fault has been relatively quiet, although in 1957 a m_b = 5.3 magnitude shock occurred on the fault causing considerable damage in Daly City. No evidence of tectonic creep (aseismic slip) has been found on this segment of the fault. There is a vast literature on the 1906 San Francisco earthquake, with the most comprehensive report published by the California State Commission (Ref. 102). Richter (Ref. 2) also discusses the 1906 earthquake and other major earthquakes of California, with many references to original data and reports of the early earthquakes on the San Andreas fault. North of San Jose, a group of faults that are subsidiary elements of the San Andreas system have produced large earthquakes, notably on the Hayward fault in the east San Francisco Bay area in 1836 and 1868. Creep on the Hayward fault in the east bay area averages 0.4 - 0.6 cm/year. South of San Jose, the faults of this group have characteristics similar to the San Andreas fault south of Los Gatos.

Between Los Gatos and Bitterwater Valley, the San Andreas fault has been very active but no great earthquakes have occurred on this segment. Creep was first discovered on this segment of the fault at the Taylor (Almaden) winery near Hollister. The average

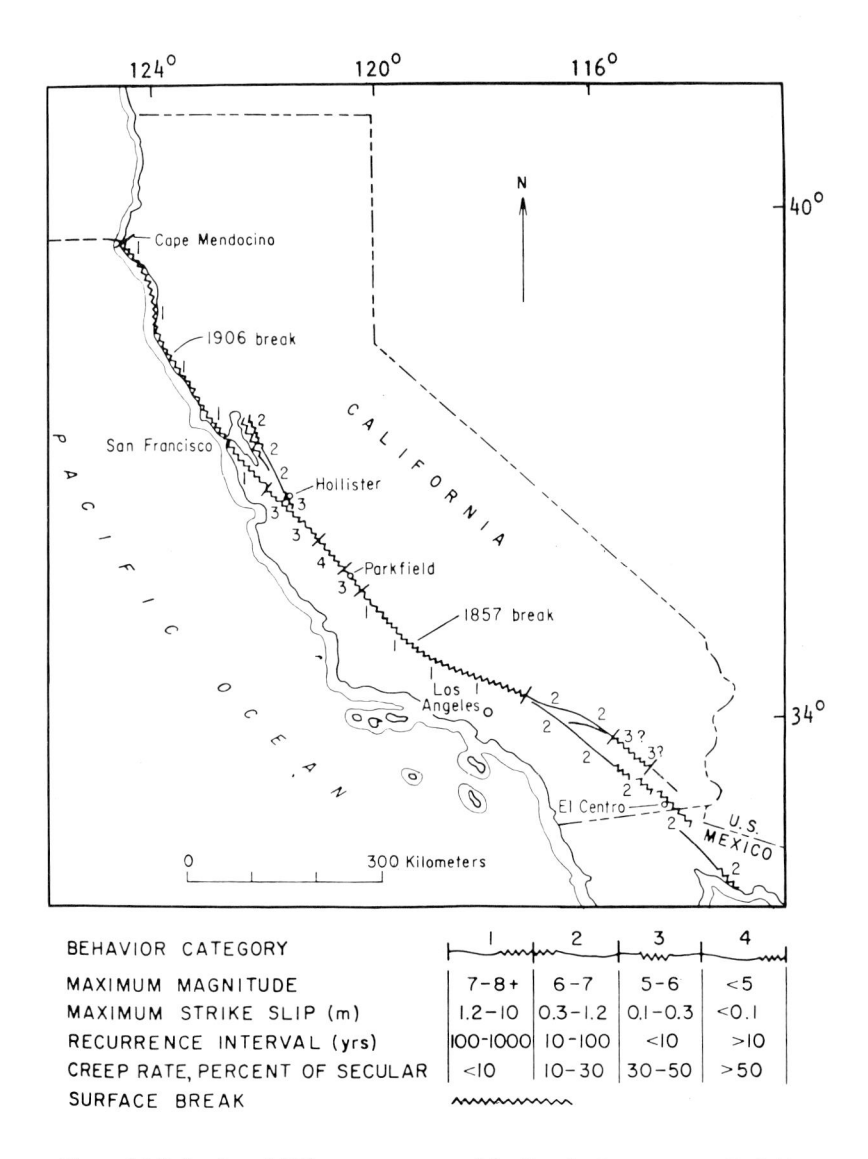

BEHAVIOR CATEGORY	1	2	3	4
MAXIMUM MAGNITUDE	7-8+	6-7	5-6	<5
MAXIMUM STRIKE SLIP (m)	1.2-10	0.3-1.2	0.1-0.3	<0.1
RECURRENCE INTERVAL (yrs)	100-1000	10-100	<10	>10
CREEP RATE, PERCENT OF SECULAR	<10	10-30	30-50	>50
SURFACE BREAK				

Figure 26. Behavior of different segments of the San Andreas system (Ref. 101).

creep rate is about one cm per year. South of Bitterwater Valley to Parkfield, the San Andreas fault has high creep rates (3 cm/year) and moderate, although occasionally damaging, earthquakes. In the Parkfield-Cholame area, earthquakes with magnitudes of 5.5 to 6.5 (M_L) occurred in 1901, 1922, 1934, and 1966.

On May 2, 1983, a magnitude m_b = 6.2 (M_s = 6.5) earthquake occurred about 14 km N22°E of Coalinga, California (Ref. 12). The central commercial district was heavily damaged, as were older residences in the town. The earthquake was not on a fault of the San Andreas system. There is, at present, no surface fault rupture known to be associated with the mainshock but surface displacement has been associated with an aftershock on June 11. This earthquake is the most damaging in California since the 1971 San Fernando shock.

The segment of the fault between Cholame and Cajon Pass ruptured during the great 1857 shock. Agnew and Sieh (Ref. 104) investigated the pattern of shaking, extent of faulting, and other aspects of this earthquake. Sieh (Ref. 105) has concluded that at least nine large earthquakes have occurred along the south-central part of the San Andreas fault since the sixth century A.D. Sieh (Ref. 106) has also found historical accounts of foreshock activity on the San Andreas fault in the Parkfield-Cholame area northwest of the fault rupture of the 1857 earthquake. If small earthquakes on a segment of the San Andreas fault known for fault creep "triggered" the 1857 earthquake, it implies that the 1857 rupture broke from north to south. If the probability of fault rupture in a particular direction could be determined in advance of a future large earthquake, this information could be used to refine any estimation of the spectrum and duration of shaking of the resulting motion in areas likely to experience damaging ground motion. Figure 27 is an isoseismal map developed by Agnew and Sieh for the 1857 earthquake, but it also shows the area over which the 1906 earthquake was felt. Even taking into consideration the quality and limited number of observations available for the 1857 earthquake, the felt area of the 1906 shock appears to be somewhat larger than the 1857 earthquake.

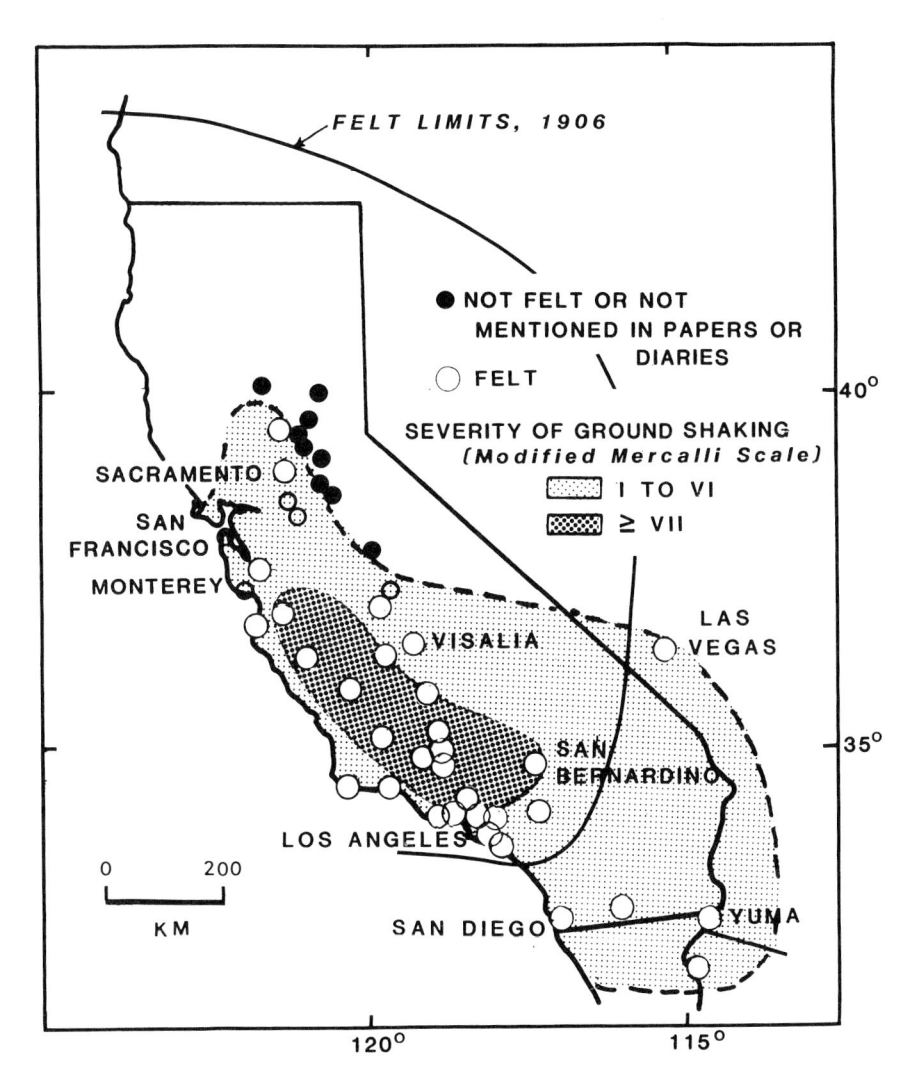

Figure 27. Isoseismal map for the January 9, 1857 earthquake on the San Andreas fault near Fort Tejon (Ref. 103). Also shown, for comparison, are the felt limits for the 1906 San Francisco earthquake.

Fault creep has not been discovered on the San Andreas along the segment that ruptured in 1857. Southeast of Cajon Pass, the San Andreas fault system consists of two main branches. The southwest branch terminates at the east trending Banning fault.

The southeast branch continues southward along the east side of the Salton Sea where fault creep has not been discovered. Earthquake activity is moderate and large shocks have not occurred historically.

Faults of the San Jacinto system, including the Imperial, Superstition Hills and Coyote Creek faults, extend from the vicinity of San Bernardino southward into Mexico. A number of earthquakes with magnitudes as large as $M_s = 7.0$ have occurred on these faults. Examples are the April 9, 1968 Borrego Mountain earthquake on Coyote Creek fault (Ref. 107) in the northwestern Imperial Valley ($M_L = 6.4$) and the numerous damaging shocks that have occurred on the Imperial fault in the south central Imperial Valley near El Centro and in northern Mexico. Virtually all of the larger shocks that have occurred on faults that make up the San Jacinto system have been associated with surface fault rupture. The October 15, 1979, $M_s = 6.6$ earthquake near El Centro again ruptured a segment of the Imperial fault that has moved repeatedly in the past, notably in 1940. The 1979 shock was well recorded on strong motion accelerographs and these data have contributed substantially to our understanding of strong ground motion (Ref. 108.)

In a review of the seismicity of Southern California for the period 1934-1963 and its relation to geologic structure, Allen et al (Ref. 109) conclude that most and perhaps all of the large historic earthquakes that have been associated with surficial fault displacements occurred on major throughgoing faults having a previous history of extensive Quaternary displacements. While all areas of high seismicity in the region they study fall within areas having numerous Quaternary fault scarps, not all intensely faulted areas were active during the 30-year period they investigated. Using the distribution of Quaternary faults as the criteria for hazard evaluation leads to a problem in short-term earthquake hazard evaluation, since in California the total area over which Quaternary faulting has occurred is always much larger than the area over which earthquake activity is likely to occur in any relatively short period of time.

Some generalizations are possible concerning the characteristics of seismicity in California and Nevada: (1) the earthquakes are nearly all shallow, usually less than 15 km in depth; (2) focal mechanisms of earthquakes in California are generally consistent with the right lateral movement of the Pacific plate with respect to the North American plate; (3) the recurrence rate for a large ($M_s > 7.8$) earthquake on the San Andreas fault system is of the order of 100 years; and (4) recurrence rates for large earthquakes on single fault segments in the Nevada seismic zone are believed to be of the order of thousands of years.

Washington and Oregon

The level of seismicity from north of Cape Mendocino to slightly south of Puget Sound, Washington is low considering the active volcanism of the Cascade Range; the physiological features of this area are given generally in Fig. 28. A number of damaging earthquakes have occurred in the Puget Sound depression, and one large earthquake is known to have occurred east of the Cascades in 1872. The historical seismicity is shown in Fig. 29 and listed in Table 12. Table 13 lists the distribution of earthquakes by maximum intensity in Washington and Oregon.

As previously noted, the primary cause of the seismicity of the entire western United States is the differential motion between the North American and Pacific plates. Figure 30 schematically illustrates the principal tectonic elements that are considered for the Washington-Oregon region. In general the San Andreas fault and the Queen Charlotte Islands-Fairweather fault can be considered as related transform faults connecting the east Pacific rise (where oceanic crust is generated) and the Aleutian trench (where oceanic crust is consumed). This large fault system would represent a line of pure slip between the North American and Pacific plates, provided that these plates behaved in a rigid manner (Ref. 110). An exception to rigid-plate motion exists for the shaded zone of Fig. 30 where plate segments are being internally deformed. An improved model of the interaction of these subplates with the North American plate should lead to a much better understanding of the seismicity of northern California, Oregon, and Washington.

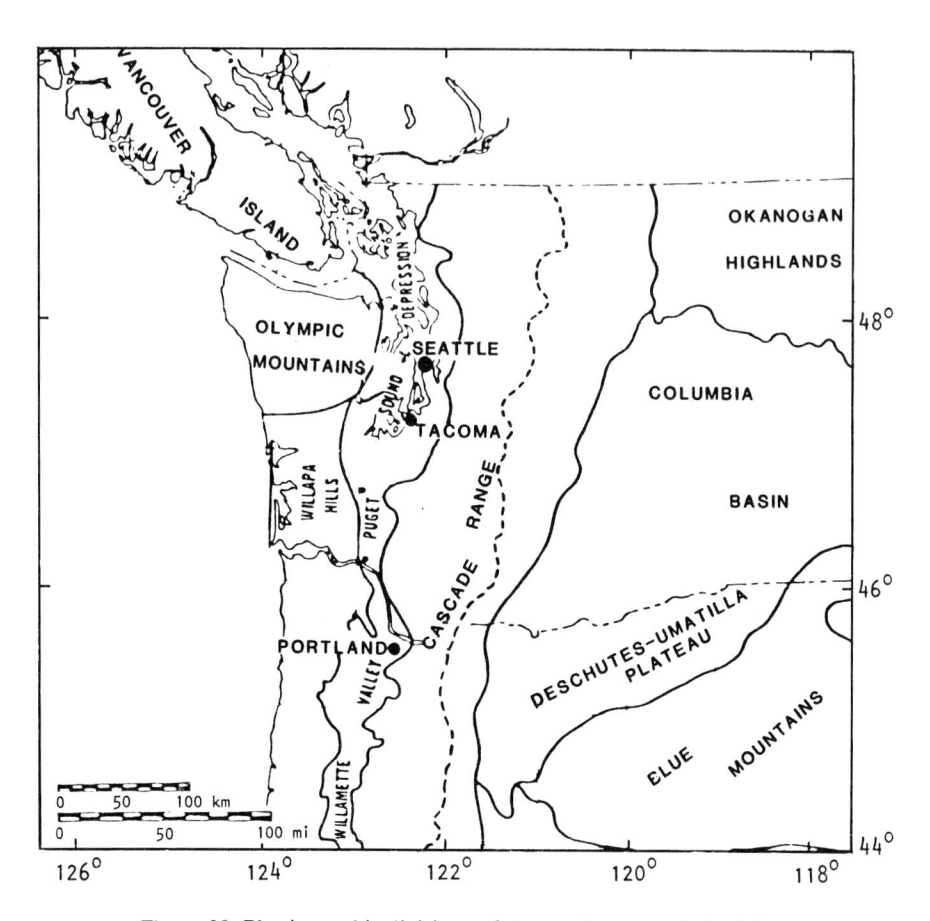

Figure 28. Physiographic divisions of the northwestern United States.

Important questions are whether the Juan de Fuca plate is presently being actively subducted beneath the North American plate and what is the origin of the large, damaging subcrustal earthquakes that have occurred in the Puget Sound area. Geophysical, stratigraphic, or tectonic arguments supporting present-day subduction in the northwest have been provided by Riddihough (Refs. 113, 114), Riddihough and Hyndman (Ref. 115), Kulm and Fowler (Ref. 116), and Atwater (Ref. 117), among others; however, other seismological (Refs. 93, 118) and tectonic evidence (Ref. 119) has been used to argue against subduction.

The two most recent, damaging earthquakes in the Puget Sound area (M_s = 6.5, 1965; M_s = 7.1, 1949) are known from reasonably well constrained instrumental data to have occurred at depths of about 60 km and 70 km, respectively (Fig. 31). Based

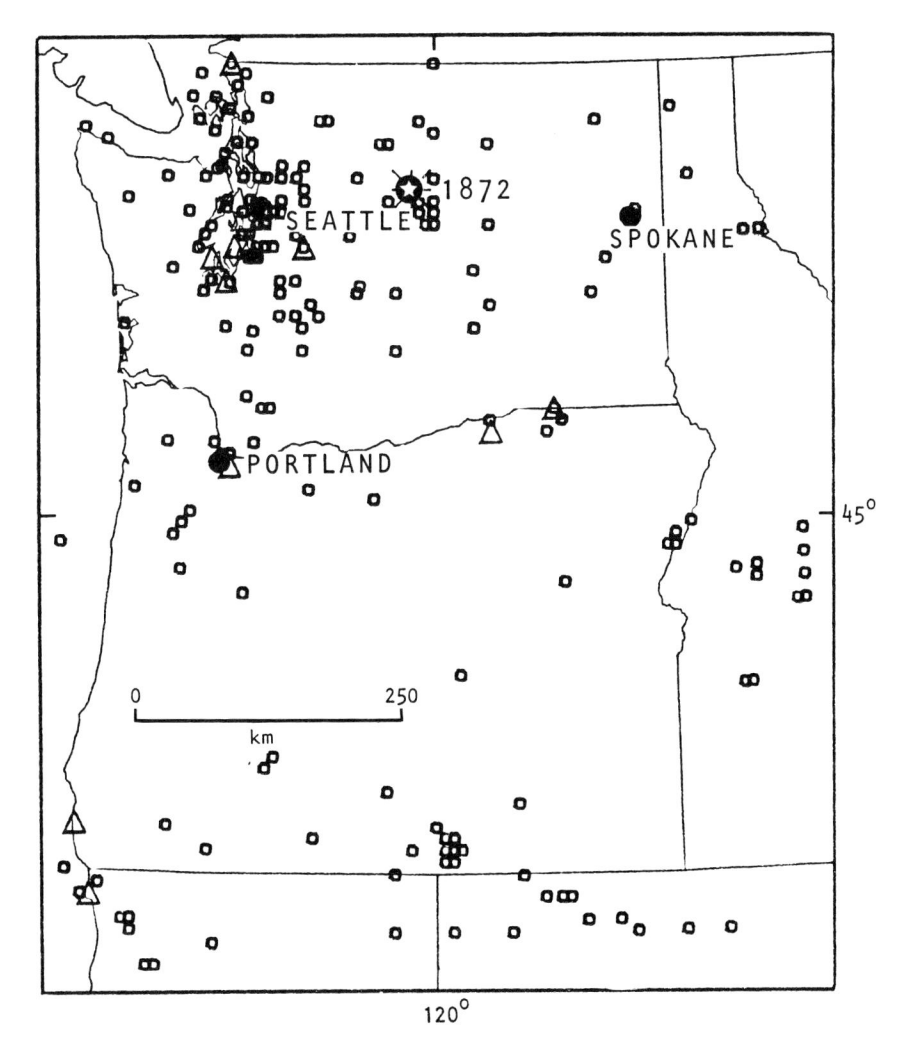

Figure 29. Seismicity of Oregon and Washington, 1859-1975. The star represents an earthquake with maximum Modified Mercalli intensity of IX, triangles represent earthquakes with maximum intensities of VII-VIII, and small squares represent earthquakes with maximum intensities of V-VI.

Table 12

*Important Earthquakes of
Washington and Oregon**

Date (GMT)	Location	Maximum MM Intensity (I_o)	Magnitude M_s
Dec 14, 1872	Near Lake Chelan, WA (Probably shallow depth of focus)	IX	(7.0)
Oct 12, 1877	Cascade Mountains, OR	VIII	
Mar 7, 1893	Umatilla, OR	VII	
Mar 17, 1904	About 60 km NW of Seattle	VII	
Jan 11, 1909	North of Seattle, near Washington/British Columbia	VII	
Dec 6, 1918	Vancouver Island, B.C.	(VIII)	7.0
Jan 24, 1920	Straits of Georgia	(VII)	
Jul 16, 1936	Northern Oregon, near Freewater	VII	(5.7)
Nov 13, 1939	NW of Olympia (Depth of focus about 40 km)	VII	(5.8)
Apr 29, 1945	About 50 km SE of Seattle	VII	
Feb 15, 1946	About 35 km NNE of Tacoma (Depth of focus 40-60 km)	VII	6.3
Jun 23, 1946	Vancouver Island	(VIII)	7.2
Apr 13, 1949	Between Olympia and Tacoma (Depth of focus about 70 km)	VIII	7.1
Apr 29, 1965	Between Tacoma and Seattle (Depth of focus about 59 km)	VIII	6.5

*Source: U.S. Geological Survey Data File and Ref. 111, except for the 1872 event (Ref. 129) and the June, 1946 event (Ref. 112). Parentheses indicate either maximum intensity is uncertain or magnitude has been estimated from intensity data.

on the observation that these two subcrustal earthquakes were characterized by an extremely low level of aftershock activity, and noting that the 1939 and 1946 earthquakes also exhibited low aftershock activity, Algermissen and Harding (Ref. 120) concluded that the damaging 1939 and 1946 shocks were also subcrustal. Such a lack of aftershock activity is characteristic of subcrustal earthquakes. A possible physical basis for the lack of aftershock activity for subcrustal earthquakes appears to be related to the small source dimensions of these earthquakes and to the high hydrostatic pressure that exists at depths, which has the effect of quickly arresting any rupture process (Ref. 121). There is some evidence that the April 29, 1945 earthquake located about 50 km southeast of Seattle was a shallow event. The main shock was followed by a number of moderately strong aftershocks. Interpretation of data obtained from the University of Washington seismograph network installed in 1970 in Puget Sound region shows that, while a few earthquakes occur at depths of up to 80 km in the Puget Sound area, nearly all of the low level seismicity recorded by the network occurs at depths less than about 30 km (Ref. 118).

It should be noted that a number of large earthquakes have occurred northwest of Puget Sound in the vicinity of Vancouver Island and northwestward to the Queen Charlotte Islands. Reviews of seismicity and earthquake risk in these areas have been presented by Milne et al (Ref. 122) and Basham et al (Ref. 123). Large earthquakes at this distance from Washington have, in the past, caused only minor damage. However, increasing urbanization of the Puget Sound area may increase the potential for damage to tall structures from long period ground motion.

The largest historical shallow earthquake in Washington and Oregon is probably the 1872 shock, centered east of the Cascades in the vicinity of Lake Chelan. Historical accounts of the effects

Table 13

Distribution of Earthquakes by Maximum Intensity in Washington and Oregon

Maximum MM Intensity	Number
V	150
VI	57
VII	8
VIII	3
IX	1

*Source: U.S. Geological Survey Data File and Refs. 111 and and 112.

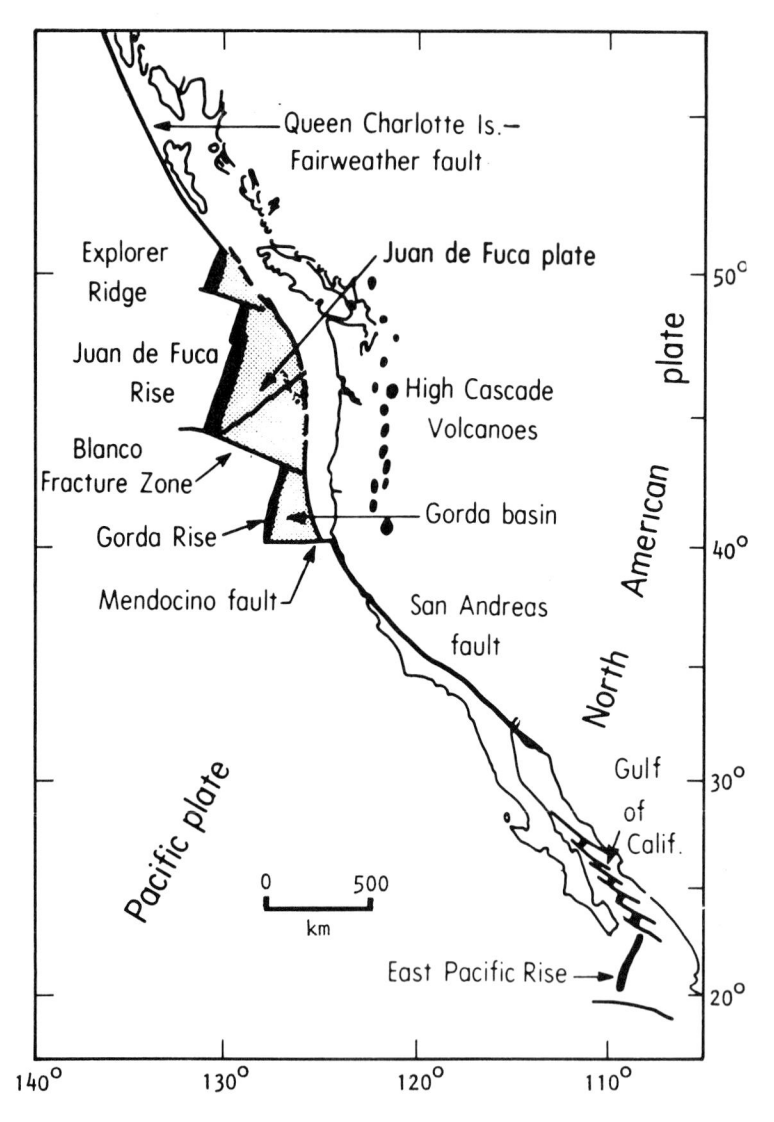

Figure 30. Tectonic elements for the northeast Pacific and western North America (adapted from Ref. 112).

of this earthquake were investigated extensively by Dale during the preparation of the Preliminary Site Analysis Report (PSAR, Docket 50522) for a proposed Skagit nuclear power plant in northern Washington (Ref. 7). Since her initial study, the available information on the earthquake has been reviewed by a number of groups (Refs. 124, 125, 126, 127, 128). The most recent review (Ref. 129) suggests that the epicenter of the 1872 event occurred in the vicinity of Lake Chelan and had a maximum MM intensity of IX and a magnitude of the order of $M_s = 7.0$, and probably a shallow depth of focus. The presumed shallow focal depth is based on the historical accounts of a long aftershock sequence that followed the main shock.

The occurrence of such a large, shallow earthquake in central Washington, the known occurrence of large ($M_s \approx 7.0$) earthquakes with foci at depth of 60-70 km in the Puget Sound area, the known distribution of shallow seismicity of moderate magnitude in Puget Sound, and evidence of Holocene surface

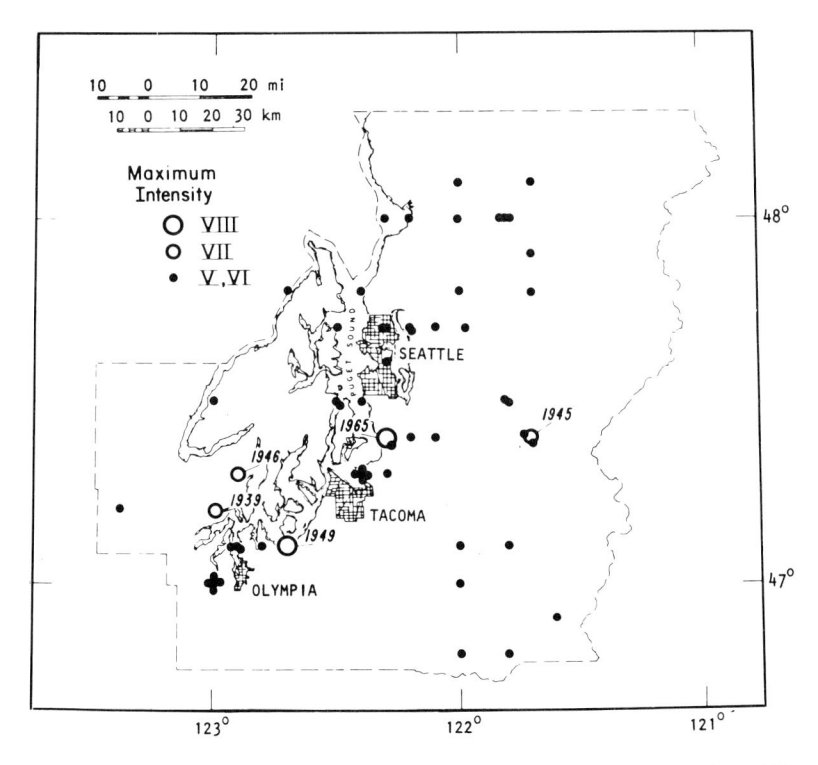

Figure 31. Seismicity in the six-county Puget Sound area for the period 1859-1975, only intensities ≥ 5 (Ref. 111).

ruptures west of Puget Sound (Ref. 130), all raise the question of the possibility of large shallow earthquakes in the Puget Sound area. No unifying, proven regional tectonic model for the occurrence of earthquakes in the Puget Sound area has yet been developed, and consequently the future occurrence of such large shallow shocks must be considered.

Elsewhere in Washington and Oregon the seismicity is sporadically distributed and of only moderate magnitude. In northeastern Oregon and southeastern Washington there is a strong northwest trending structural control of the geologic features, the most prominent of which are the Olympic-Wallowa lineament (Ref. 131) and the Vail Zone (Ref. 132). The control of these northwest trending zones on the regional distribution of seismicity is not well understood. The largest earthquake known to have occurred in this area is the July 16, 1936 northern Oregon shock (M_s = 5.7, I_o = VII) that caused damage at Freewater, State Line and Umapine. Aftershocks were felt for several months.

Research during the past ten years that has led to the reassessment of the importance of the December 14, 1872 northcentral Washington earthquake and the identification of Holocene faulting on the west side of Puget Sound have forced a reassessment of the possibility of large shallow shock not only in the Puget Sound area but east of the Cascades. With the exception of the 1872 earthquake, the seismicity outside of the Puget Sound depression has been low to moderate. Evaluation of seismic hazard has been hampered by the lack of clear relationships between seismicity and geologic structure except in the very general sense of plate interaction between the North American and Pacific plates.

ALASKA

The Alaska-Aleutian Island area is one of the most active seismic zones in the world. In southeast Alaska, the Pacific plate slides past the North American with right-lateral strike slip movement along the Queen Charlotte Island-Fairweather fault system (Fig. 30). The Pacific plate is bounded on the north by the Aleutian Trench where it begins to be subducted beneath the North American plate. Figure 32 shows some of the major faults of southcentral Alaska, and Figs. 33 and 34 show the historical seismicity of Alaska and the Aleutian Islands. Table 14 lists major earthquakes, and Table 15 gives the magnitude distribution of earthquakes in Alaska and the Aleutians. It is clear from Figs. 33 and 34 that the entire coastal region of Alaska and the Aleutians have experienced great earthquake activity even in the short time period (1899-present) for which the seismicity is reasonably well known. Even though the seismicity extends in a band along the southern coastal and offshore areas, the density of large earthquakes along the island arc has not been uniform, which emphasizes the problem that the seismic history of Alaska is known for such a relatively short time. The earliest permanent settlements were established on Kodiak Island in 1783 and on Cook Inlet in 1788. The pattern of establishing coastal settlements in Alaska changed little until the discovery of gold in the 1890's, when population growth turned inland. The increase in population and its dispersal throughout Alaska have resulted in a greater number of reported earthquakes. Probably no great earthquakes have been omitted since the turn of the century, but only in the past 20-25 years has the seismic history been reasonably well known down to the level of $M_s = 5.0$.

Magnitude was used in preparing Tables 14 and 15 since the low density of population and geography of Alaska guarantees the incompleteness of intensity data even for the largest shocks.

Figures 33 and 34 indicate that, given a sufficiently long seismicity record, the distribution of large earthquakes would

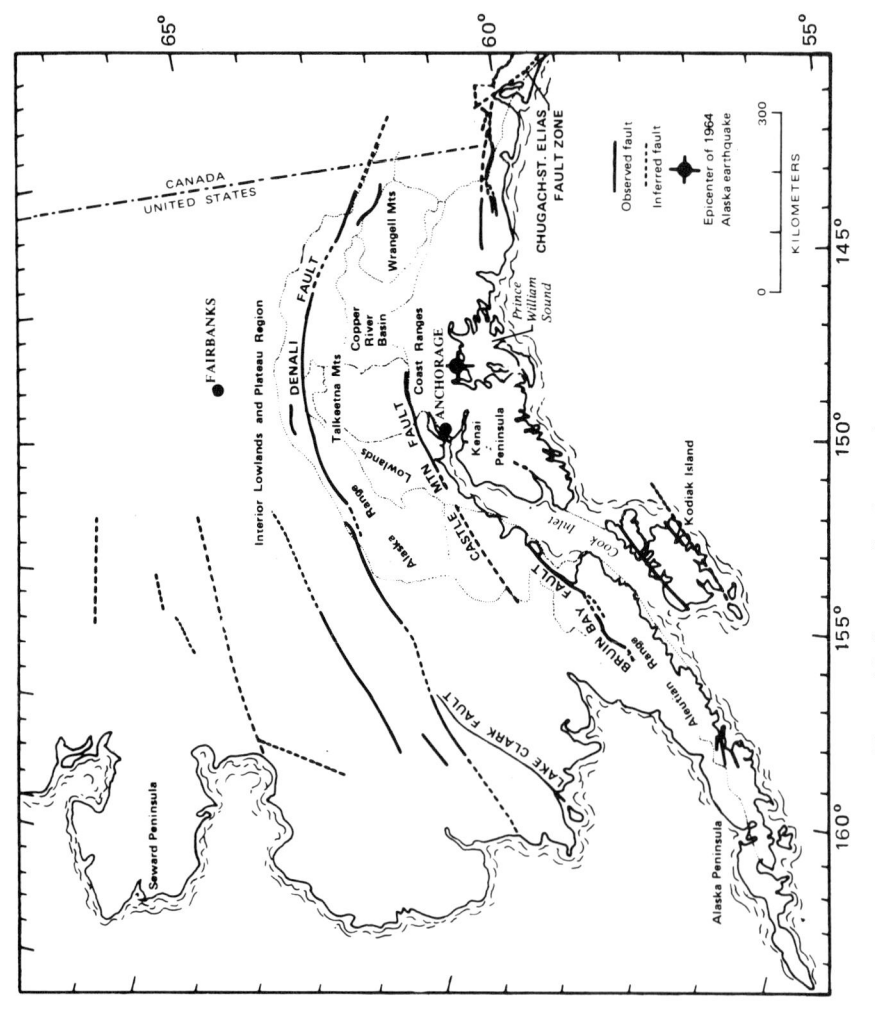

Figure 32. Some major faults of central Alaska.

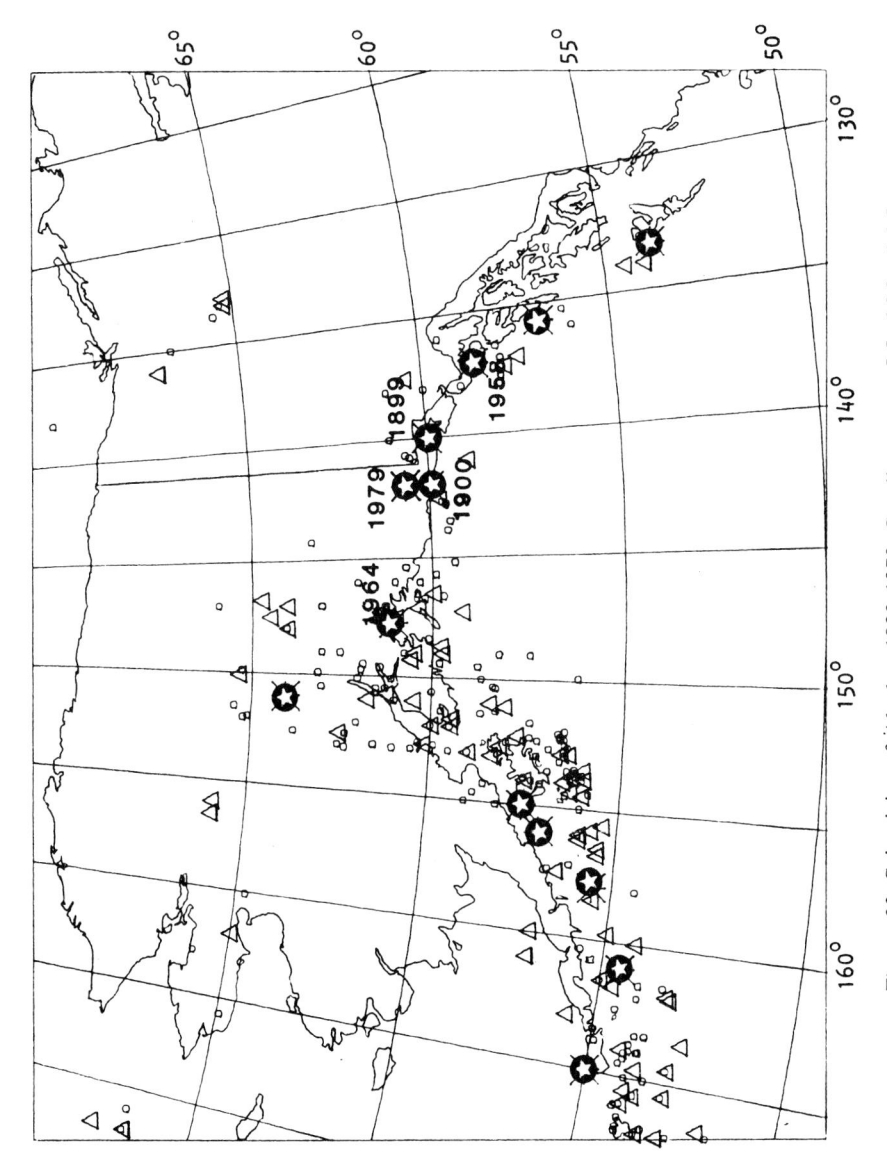

Figure 33. Seismicity of Alaska, 1899-1979. Small squares, $5.5 \leq M_s < 6.5$; triangles, $6.5 \leq M_s < 7.5$; stars $M_s \geq 7.5$.

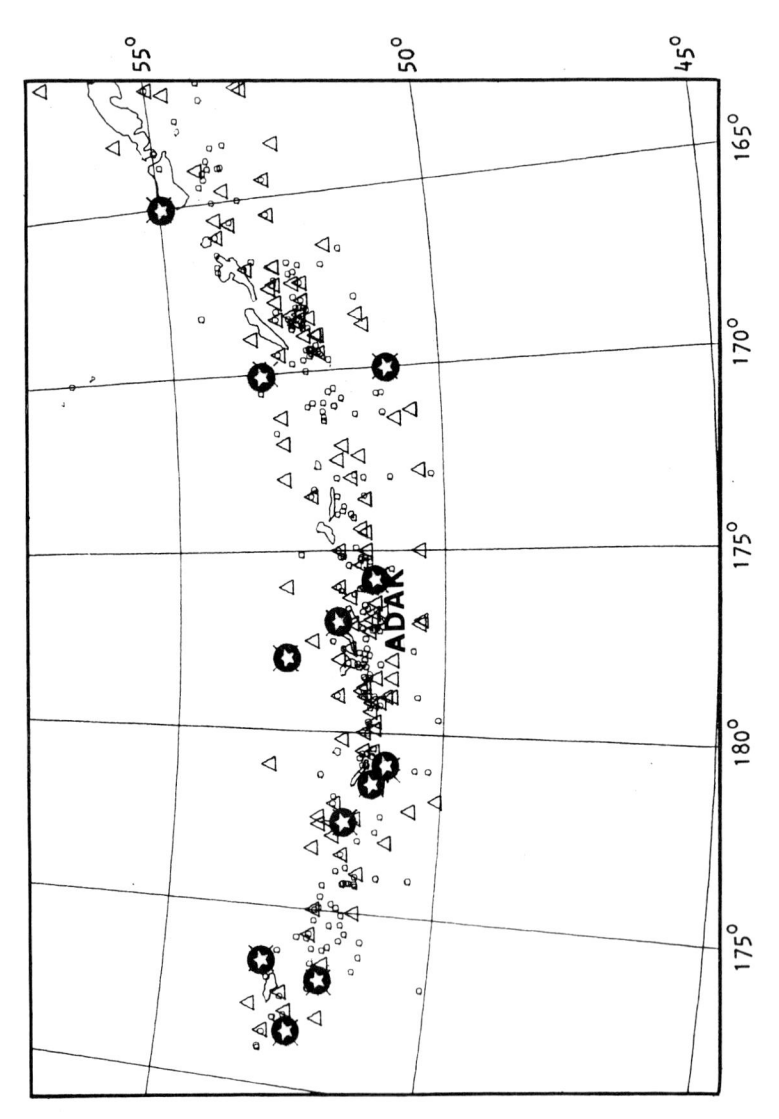

Figure 34. Seismicity of the Aleutian Islands, 1899-1979. Small squares, $5.5 \leq M_s < 6.5$; triangles, $6.5 \leq M_s < 7.5$; stars $M_s \geq 7.5$.

Table 14

*Major Earthquakes of Alaska**

Date (GMT)	Location	Magnitude M_s
Sep 4, 1899	Near Cape Yakataga	8.3
Sep 10, 1899	Yakutat Bay	8.6
Oct 9, 1900	Near Cape Yakataga***	8.3
Jun 2, 1903	Shelikof Straight	8.3
Aug 27, 1904	Near Rampart	8.3
Aug 17, 1906	Near Amchitka Island	8.3
Mar 7, 1929	Near Dutch Harbor	8.6
Nov 10, 1938	East of Shumagin Islands	8.7
Aug 22, 1949	Queen Charlotte Islands (Canada)	8.1
Mar 9, 1957	Andreanof Islands	8.2
Mar 28, 1964	Prince William Sound	8.4**
Feb 4, 1965	Rat Islands	7.8**

*Source: U.S. Geological Survey data files, and Refs. 2 and 183.
**These earthquakes have much larger moment magnitudes. See Appendix B.
***Stover et al (Ref. 184) have suggested that this earthquake occurred near Kodiak Island, 600 km to the west.

Table 15

Distribution of Earthquakes by Magnitude in Alaska (1899-1976)

M_s	Number
5.0 - 5.9	757
6.0 - 6.9	344
7.0 - 7.9	63
≥ 8.0	11

eventually become nearly uniform along the island arc. This is a reasonable expectation and the basis for the idea of a seismic gap. Figure 35 shows the aftershock zones of earthquakes with magnitudes of 7.3 or greater since 1938 along the Pacific coast of Alaska. The aftershocks that occur after large shallow earthquakes usually define the zone that ruptures during the main

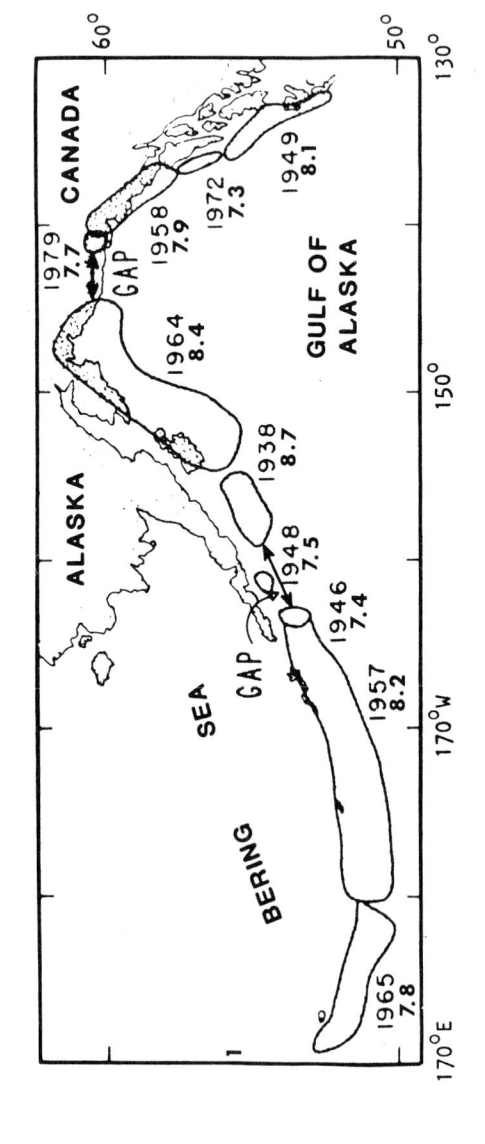

Figure 35. Aftershock zones of earthquakes of magnitude M_s = 7.3 or greater since 1938 along the Pacific Coast of Alaska (adapted from Ref. 133). The year of occurrence and magnitude of selected earthquakes are indicated.

shock. Note that two pronounced seismic gaps can be seen in Fig. 35: one between the aftershock zones of the 1979 (M_s = 7.7) St. Elias earthquake and the 1964 Prince William Sound earthquake, called the Yakataga gap; the other between the aftershock zones of the 1938 (M_s = 8.7) and 1946 (M_s = 7.4) shocks, known as the Shumagin gap. The seismological evidence for the Shumagin gap has recently been extensively reviewed (Ref. 134). A tectonic model (Ref. 135) has been proposed for southeast Alaska in the transition zone (from the Chugach-St. Elias fault zone southward, Fig. 32) between strike-slip faulting on the Queen Charlotte-Fairweather fault to thrusting of the Pacific plate under the North American plate along the southern coast of central Alaska. The Pacific plate is moving northward with respect to the North American plate along the Queen Charlotte-Fairweather fault (Fig. 30) at the rate of about 6 cm/year, but in addition there is a small amount of convergence of the plates (about 1 cm/yr) which results in minor subduction of the oceanic plate beneath the continental margin. According to this idea, the effects of this convergence are uplift of the Chugach-St. Elias range and deformation at the Denali fault in the central Alaska Range.

Large earthquakes in the Aleutian arc and southeast Alaska often occur without great life or property loss due to vibration damage because of the sparse population in many parts of Alaska. A good example is the small amount of damage resulting from the 1979 St. Elias earthquake (M_s = 7.7). However, large earthquakes ($M_s > 7.7$) always represent a serious potential tsunami threat, not only to Alaska but to Hawaii, other islands of the Pacific and the coastal area surrounding the Pacific. The St. Elias earthquake has been described in a set of papers in the Oct. 1980 issue of the *Bulletin of the Seismological Society of America*.

A somewhat mitigating factor in the seismic hazard in Alaska is the fact that Alaska is one of the few regions of the United States where earthquakes occur at intermediate depths. Earthquakes occur at depths to about 250 km in a zone dipping to the north under the Aleutian-Alaskan arc because of the subduction of the Pacific plate under the North American plate.

Earthquakes up to about $M_s = 6.5$ cause little damage when their focal depths are greater than about 80 km. A typical cross section is shown in Fig. 36. In addition to the earthquakes that occur with intermediate focal depths, shallow earthquakes do occur along the Aleutian arc and in southcentral and central Alaska. There was considerable damage at Fairbanks and damage to the Alaska railroad in 1947 as a result of a $M_s = 7.0$ magnitude earthquake about 65 km southeast of the city.

The most devastating and probably the largest historical earthquake in Alaska occurred on Good Friday, March 28, 1964. The magnitude was $M_s = 8.3$ (Ref. 137). The earthquake has more

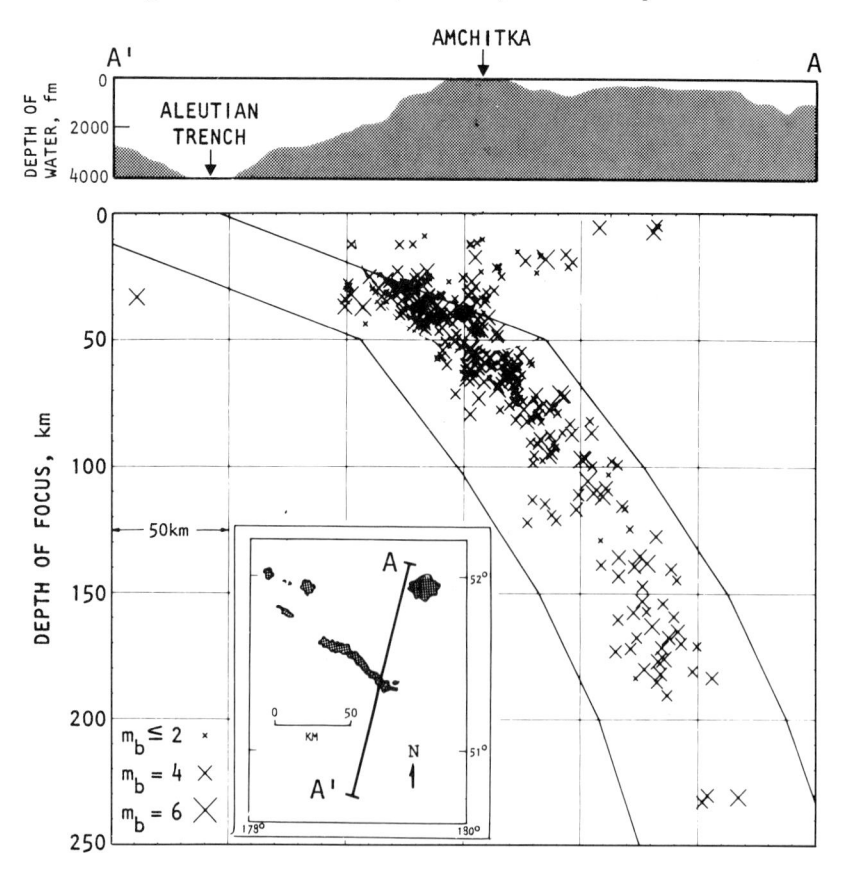

Figure 36. A typical depth profile of earthquake foci perpendicular to the Aleutian arc through Amchitka (adapted from Ref. 136).

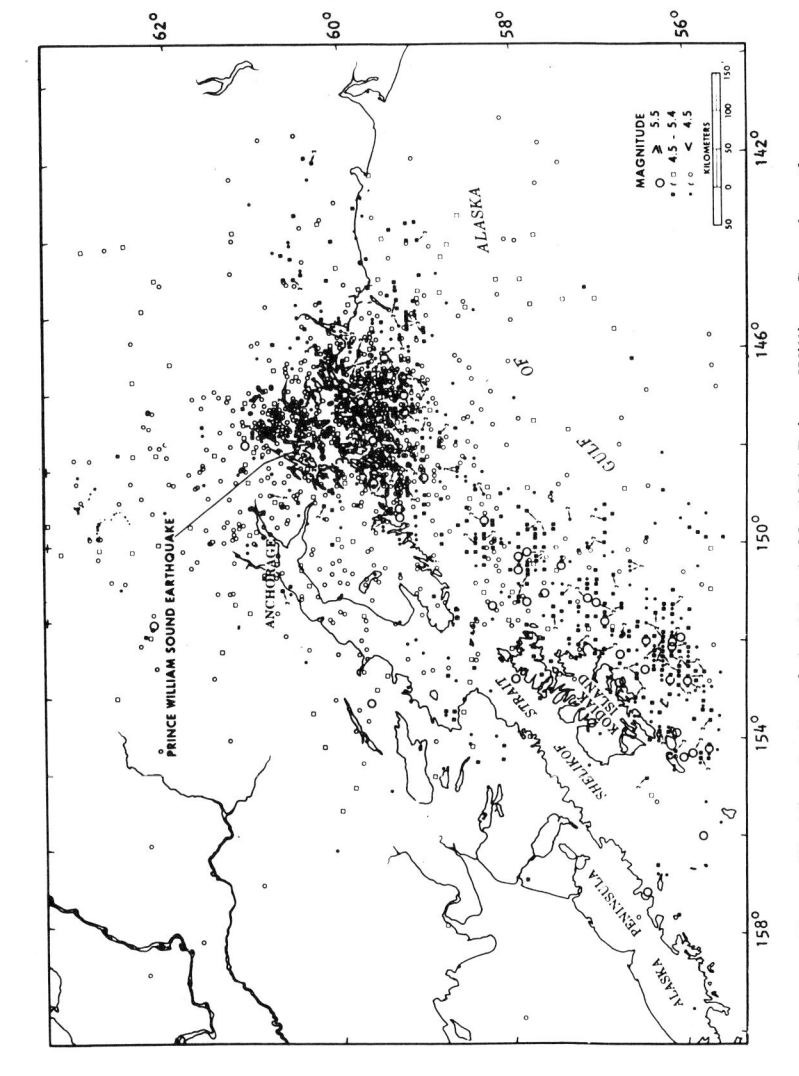

Figure 37. Aftershocks of the March 28, 1964 Prince William Sound earthquake, March 28-December 31, 1964 (Ref. 138).

recently been assigned a moment magnitude (M_w) of 9.2. About 114 lives were lost, principally as a result of the tsunami that followed the earthquake. Although the epicenter was located in Prince William Sound about 165 km southeast of Anchorage, the rupture surface was quite large, extending from Prince William Sound to south of Kodiak Island. Epicenters of aftershocks were dispersed throughout an area of 250,000 km^2, mainly along the continental margin of the Aleutian Trench between Prince William Sound and south of Kodiak Island. Figure 37 shows the aftershocks through December 31, 1964. Regional deformation was extensive, covering an area of about 280,000 km^2 from southwest of Kodiak Island northeast to Prince William Sound and turning southeast along the coast as far as Yakataga. Uplift exceeded 11 meters on Montague Island, where secondary faulting was also found. The maximum subsidence was about 2.3 meters. There were measured horizontal shifts of as much as 19.6 meters (Ref. 139).

Vibration damage was extensive in Anchorage, but it was mostly confined to buildings taller than two stories. Soil liquefaction in Anchorage caused widespread damage to both commercial buildings and dwellings. The greatest economic and life loss in coastal Alaska was caused by the resulting tsunami that propagated throughout the Pacific, and also by seiches in the various fjords and channels around the periphery of Prince William Sound. Towns were devastated along the Gulf of Alaska, and there was serious damage at Port Alberni, Canada, along the west coast of the United States and in Hawaii. The earthquake and its effects have been reported in many journals and in three principal government reports, including a special committee report of the National Academy of Sciences (Refs. 140, 141, 142).

HAWAII

The Hawaiian Islands have a known history (since 1834) of moderate seismic activity with a number of shocks causing damage in the intensity VII-VIII range. The seismicity is related to the well known volcanic activity of the islands in a complex and as yet little understood way. Nearly all of the seismicity is associated with the Island of Hawaii (Fig. 38). Important damaging earthquakes occurred in 1868, 1938, 1951, and 1975 (Table 16). The 1868 earthquake was probably the largest of these. The epicenter was southeast of the island of Hawaii but it was also strongly felt on Maui. Clocks were stopped at Honolulu. Damage was extensive even to wood dwellings on Hawaii. A number of deaths were caused by the associated tsunami. There was fissuring near the epicentral region and volcanic eruptions about five days after the main shock. A foreshock sequence preceded the main shock (Ref. 18). The epicenter of the 1938 earthquake was north of the island of Maui, with an estimated $100,000 damage on the island. The 1951 earthquake was near Kona where it caused widespread damage over a distance of 80 km along the coast, and a small tsunami was generated. The November 29, 1975 earthquake, with its epicenter well located instrumentally on the south flank of Kilauea volcano near the southeast coast of Hawaii, caused significant damages. The magnitude was M_s = 7.2, making this the largest earthquake in the immediate area since 1868. The source mechanism was interpreted by Wyss et al (Ref. 143) as dip slip normal faulting on a plane dipping 20° to the southeast. It has been suggested that magma injected into rift zones since the 1868 earthquake may have steadily increased the compressional stress, which was released in the 1975 earthquake (Ref. 144).

The 1975 earthquake caused extensive damage in the southeastern portion of the island of Hawaii, with two fatalities and estimated losses of $4.1 million. The accompanying tsunami had a height of 6 meters near the epicenter. There was an

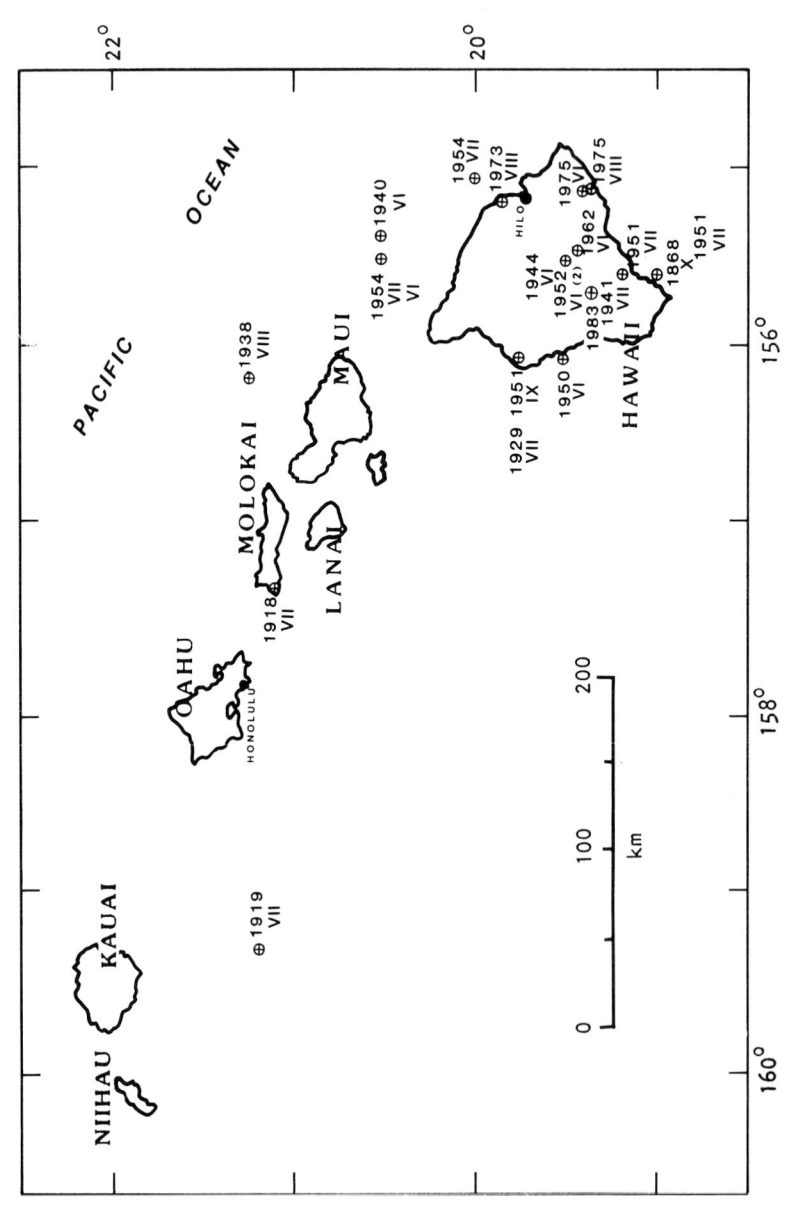

Figure 38. Seismicity of the Hawaiian Islands (Ref. 18).

extensive foreshock and aftershock sequence with a small volume volcanic eruption about 45 minutes after the main shock (Ref. 18). An earthquake (M_s = 6.6) on November 16, 1983 caused damage to homes and lifeline services on the island of Hawaii. Damages were estimated $6 million, and six people were injured (Ref. 12). The event was centered on the southwest flank of Mauna Loa volcano, about 70 miles southwest of Hilo.

Although the seismicity of Hawaii has been recorded for only about 100 years, during that time there have been a number of large shocks (Tables 16 and 17). Since there is every reason to believe that the historical level of activity will continue in the future, the continued urbanization of the islands suggests that earthquake damage and losses will increase in the future. In

Table 16

*Earthquakes Causing Significant Damage in Hawaii**

Date (GMT)	Location	Maximum MM Intensity	Magnitude M_s
Apr 2, 1868	Near south coast of Hawaii	X	
Nov 2, 1918	Mauna Loa, Hawaii	VII	
Sep 14, 1919	Kilauea, Hawaii	VII	
Sep 25, 1929	Kona, Hawaii	VII	
Sep 28, 1929	Hilo, Hawaii	VII	
Oct 5, 1929	Honualoa, Hawaii	VII	6.5
Jan 22, 1938	North of Maui	VIII	6.7
Sep 25, 1941	Mauna Loa, Hawaii	VII	6.0
Apr 22, 1951	Kilauea, Hawaii	VII	6.5
Aug 21, 1951	Kona, Hawaii	IX	6.9
Mar 30, 1954	Near Kalapana, Hawaii	VII	6.5
Mar 27, 1955	Kilauea, Hawaii	VII	
Apr 26, 1973	Near northeast coast of Hawaii	VIII	6.3
Nov 29, 1975	Near northeast coast of Hawaii	VIIi	7.2
Nov 16, 1983	Near Mauna Loa, Hawaii		6.6

*Source: Ref. 18. Intensities given are those observed on the islands.

Table 17

*Distribution of Earthquakes by Maximum Intensity in Hawaii**

Maximum MM Intensity	Number
V	56
VI	9
VII	9
VIII	3
IX	1
X	1

*Source: Ref. 18.

particular, there may be the potential for damage to high-rise buildings on Maui from long period ground motion from large earthquakes in the vicinity of the island of Hawaii and increased tsunami damage on all the islands.

PUERTO RICO

Some information concerning earthquake damage in Puerto Rico is known from as early as 1524-1528 when an old account states that the house of Ponce de Leon at Anasco was destroyed by an earthquake during this period (Ref. 145). Damaging shocks that have occurred on or near the island for which the epicenters are reasonably well known are shown in Fig. 39. Important damaging shocks are listed in Table 18 and the distribution of shocks by maximum intensity is given in Table 19. The maximum intensities are those observed on Puerto Rico, since most of the earthquakes have occurred offshore. During the past 120 years, major damaging earthquakes have occurred in 1867 and 1918. Both the 1867 and 1918 earthquakes had associated tsunamis that caused extensive damage on Puerto Rico.

It is interesting to note that the 1867 earthquake is located east of Puerto Rico in the Virgin Islands, while the epicenter of the

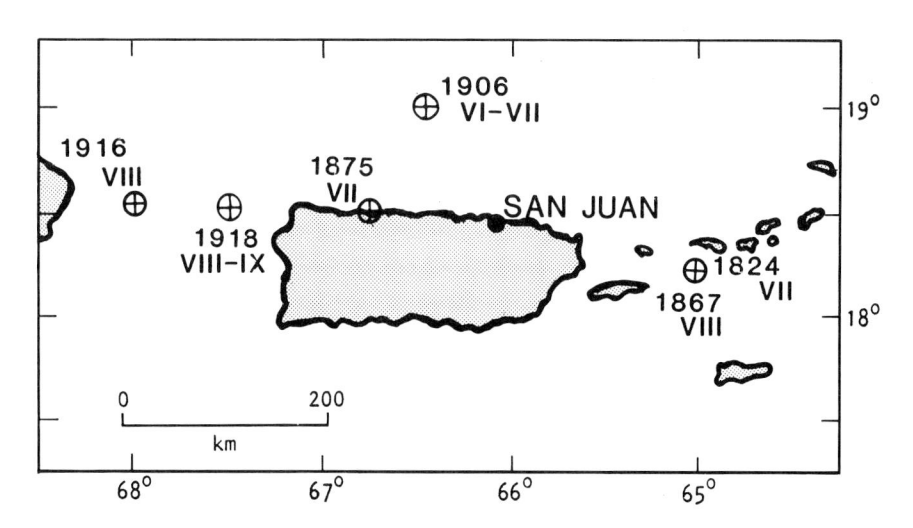

Figure 39. Earthquakes causing major damage in Puerto Rico (Refs. 18 and 145).

Table 18

Damaging Earthquakes on or Near Puerto Rico*

Date (GMT)	Location	Maximum MM Intensity	M_s
Apr 20, 1824	St. Thomas, V.I.	(VII)	
Apr 16, 1844	Probably north of Puerto Rico	VII	
Nov 28, 1846	Probably Mona Passage	VII	
Nov 18, 1867	Virgin Islands	VIII also tsunami	
Mar 17, 1868	Location uncertain	(VIII)	
Dec 8, 1875	Near Arecebo, P.R.	VII	
Sep 27, 1906	North of Puerto Rico	VI-VII	
Apr 24, 1916	Possibly Mona Passage	(VII)	
Oct 11, 1918	Mona Passage	VIII-IX also tsunami	7.5

*Intensities are from Refs. 18 and 145. Magnitudes are from Ref. 18. Parentheses mean that intensity is uncertain. Intensities given are those estimated to have occurred in Puerto Rico.

Table 19

Distribution of Earthquakes by Maximum Intensity That Have Affected Puerto Rico*

Maximum MM Intensity	Number
V	24
V-VI	4
VI	5
VI-VII	1
VII	6
VIII	2
VIII-IX	1

*Intensities are those observed on the island. Data are from Refs. 18, 19 and 145.

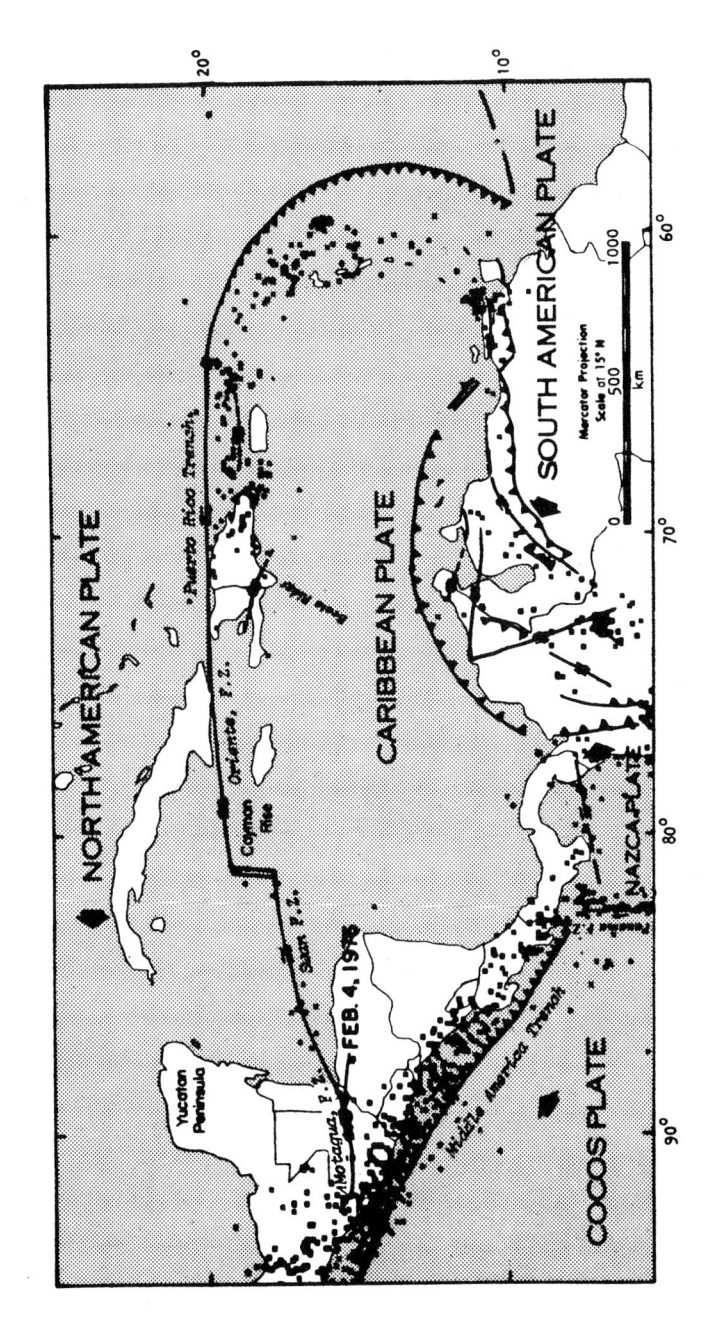

Figure 40. Tectonic setting of Puerto Rico (modified from Ref. 148). The Motagua fault zone, Swan fracture zone, Oriente fracture zone, and Puerto Rico Trench are elements of a major transform fault system on which the North American plate moves westward relative to the nearly stationary Caribbean plate. Sawtooth pattern indicates crustal subduction.

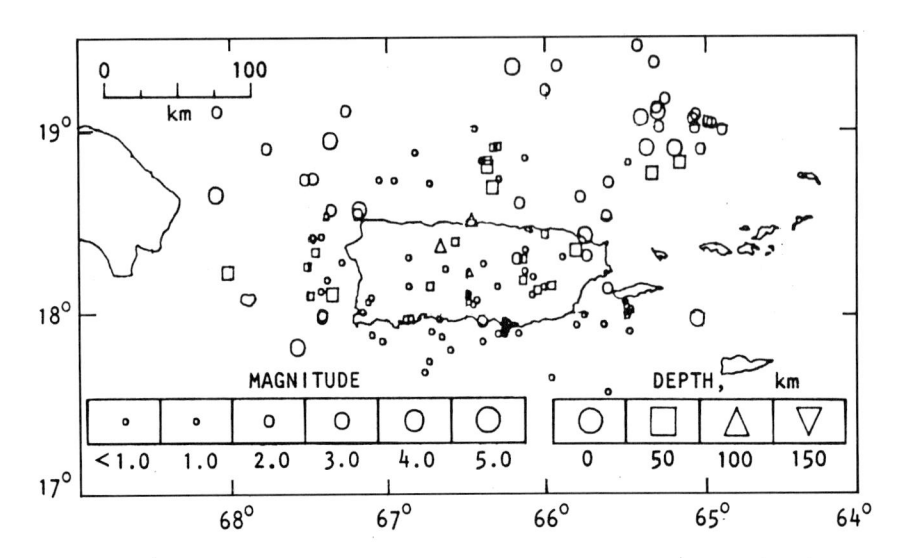

Figure 41. Puerto Rico seismicity, February 1, - May 31, 1977 (Ref. 146).

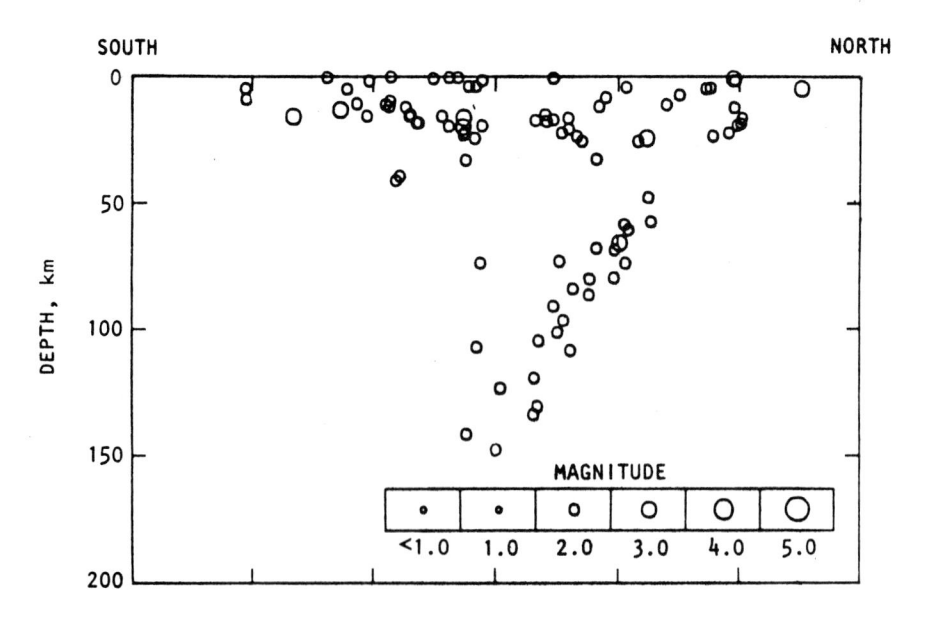

Figure 42. Depth section of Puerto Rico earthquakes, November 1, 1975-May 31, 1977. North is towards right, south towards left. Section is along longitude 66.5° W from latitude 17.0° N to latitude 20.0° N (Ref. 146).

1918 shock (M_s = 7.5) is located in Mona Passage, west of Puerto Rico. The extent of the tsunami damage and other effects of the 1918 earthquake are well documented (Ref. 147).

The tectonic setting of Puerto Rico is shown in Fig. 40. The absolute motion of the Caribbean plate is believed to be nearly fixed (Ref. 149), while the North American plate is moving westward at the rate of about 2 cm/year. Puerto Rico is located in a transition zone within the northeastern boundary of the Caribbean plate between a zone of crustal subduction in the eastern Caribbean and a series of transform faults that extends to Central America.

During 1975 the U.S. Geological Survey began the installation of a network of seismograph stations in Puerto Rico aimed at supplementing the San Juan Geophysical Observatory at Cayey that has been in existence for a number of years. The 15 station network has greatly increased the resolution of seismicity on the island. Figure 41 shows the local seismicity for the period February 1 - May 31, 1977. Figure 42 has a north-south cross section of Puerto Rico showing the seismicity for the period November 1, 1975 - May 31, 1977. Note that the low level, shallow seismicity can be easily distinguished from the seismicity associated with the subduction zone beneath the island. The amount of subduction currently going on in the zone beneath Puerto Rico is open to question, but there is clear evidence of low level seismicity in the subduction zone.

While no large earthquakes are known to have occurred on the island of Puerto Rico, the large earthquakes that have occurred east and west of the island in historic time document the fact that there is a serious tsunami hazard in Puerto Rico. In addition, there appears to be a real danger of damage to tall buildings resulting from the long period motion from potential large earthquakes offshore.

SEISMIC ZONING IN THE UNITED STATES

Introduction

The estimation of seismic hazard can take many forms and require various levels of sophistication. A simple hazard map might consist of the epicenters of known earthquakes, the maximum intensities observed or possibly the maximum intensity at every site that has reported shaking. Analogous maps can be prepared depicting fault breaks from historic, Holocene, Quaternary or other ages. Such maps imply that earthquakes will occur in the future at locations where they have occurred in the past. There are many obvious limitations to the simple approach just described. For example, the historical record of earthquakes is short, and large shocks may not have been observed in seismic source zones where the interoccurrence time of such shocks is long. In addition, active faults frequently have no surface expression, and thus geological fault slip data that could be used to extend and complement the seismological record are unavailable.

Ideally, a seismic hazard map should generalize from the seismological and geological data available and suggest where some measure of hazard, ground shaking, ground failure, surface faulting, etc., will occur in the future.

Historically, a number of different methods have been used to estimate seismic hazard in the United States. One early national zoning map, along with a detailed zoning map of Boston, appears in J. R. Freeman's book, published in 1932 (Ref. 150). Freeman also discusses earlier national zoning maps prepared by Heck, and maps of California by Willis and the Board of Fire Underwriters of the Pacific. Interest in the evaluation of seismic hazard has greatly increased in the past 20 years, at least in part because of the realization of the importance of the hazard evaluation of nuclear power plant sites and other critical facilities. Only an outline of the development of seismic zoning maps with examples of the application of these maps to building codes is attempted here. Perkins has also given a brief review of the development of zoning maps (Ref. 151).

Development of National Maps

In 1948, a "Seismic Probability Map" was developed by F. P. Ulrich of the U.S. Coast and Geodetic Survey (Ref. 152). This map divided the contiguous United States into four zones numbered 0, 1, 2, 3, where Zone 3 was considered to have the greatest potential for earthquake damage (Fig. 43). The map was adopted in 1949 by the International Conference of Building Officials (ICBO) for inclusion in the *Uniform Building Code*, and became one of the first national zoning maps used for building code purposes in the United States. The numbered zones were used in the code in the development of the lateral force provisions considered appropriate for various parts of the country. Despite the fact that Ulrich developed his map with the aid of some of the leading seismologists in the country, the exact basis for the zones on the map was never made entirely clear by Ulrich in published papers. The map displays epicenters of the larger earthquakes that occurred through 1946. The zones were apparently drawn on the basis of the maximum magnitude earthquake that had occurred in each zone. The zones are more or less geometrical in outline and do not represent differences in ground motion. Thus, at some places on the map, zone 3 adjoins zone 1, etc. Within a few years, the map was withdrawn by the U.S. Coast and Geodetic Survey as misleading and subject to misinterpretation. No map was offered as a replacement.

An important seismic regionalization map was published by Richter in 1958 (Ref. 153 and Fig. 44). This map contained several significant advances, for it depicted the estimated maximum ground motion rather than the distribution of earthquake epicenters, and it introduced the notion of frequency of occurrence of earthquakes in a qualitative way.

The 1970 edition (Ref. 154) of the *Uniform Building Code* (UBC) used a map developed by Algermissen (Ref. 155 and Fig. 45) which has the same numbering scheme (0 through 3) as the

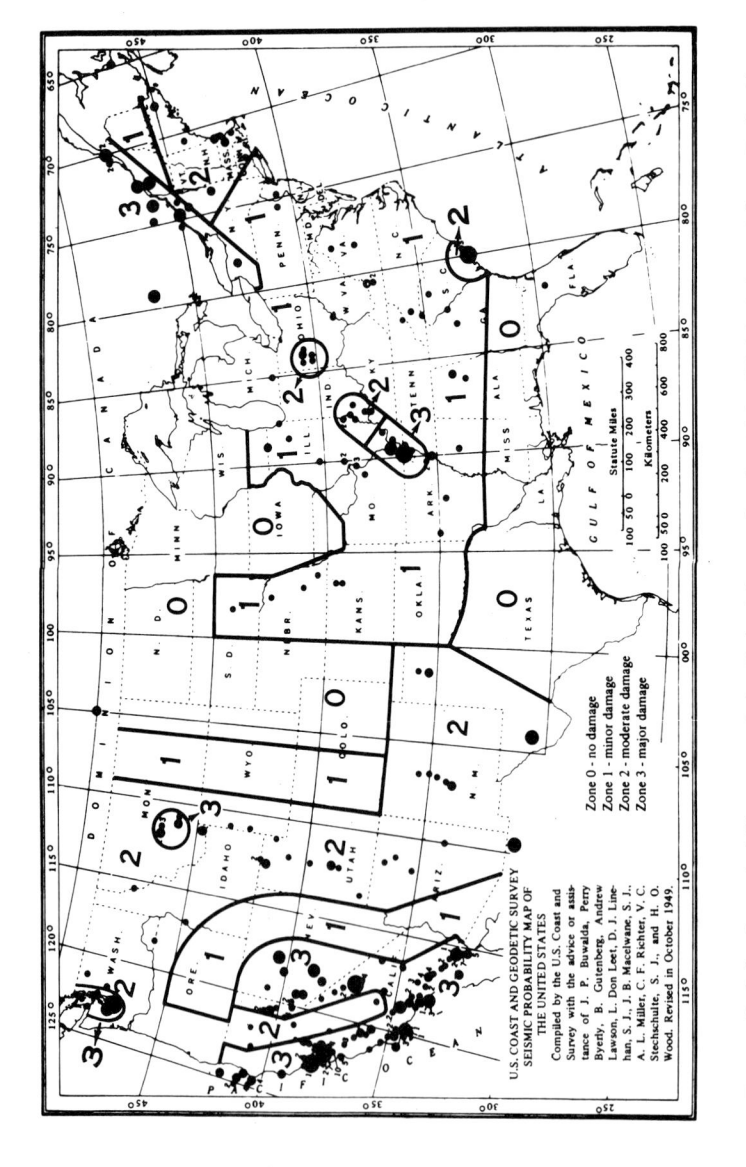

Figure 43. Seismic probability map of the United States developed by Ulrich (adapted from Ref. 152).

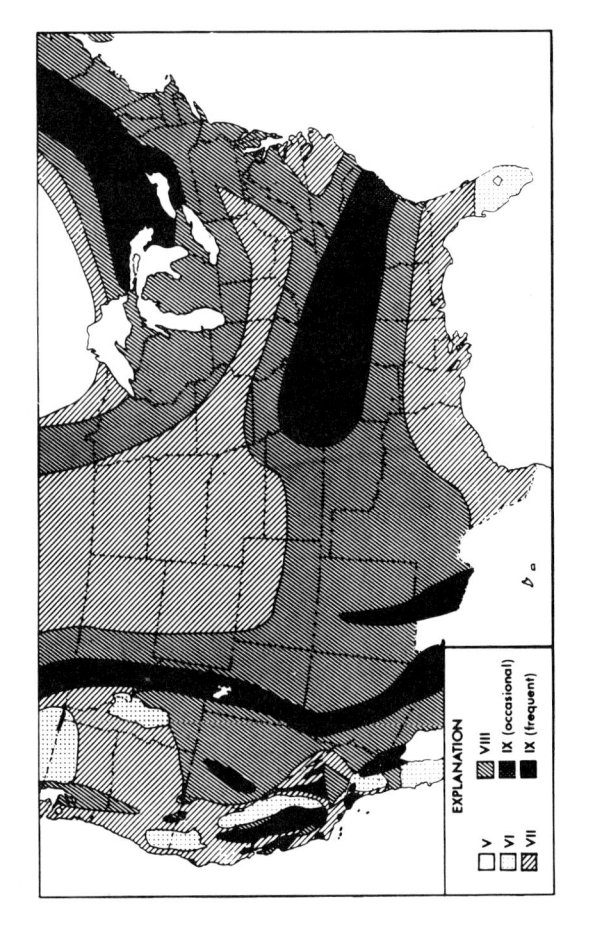

Figure 44. Seismic regionalization map of the conterminous United States published by Richter (adapted from Refs. 151 and 153).

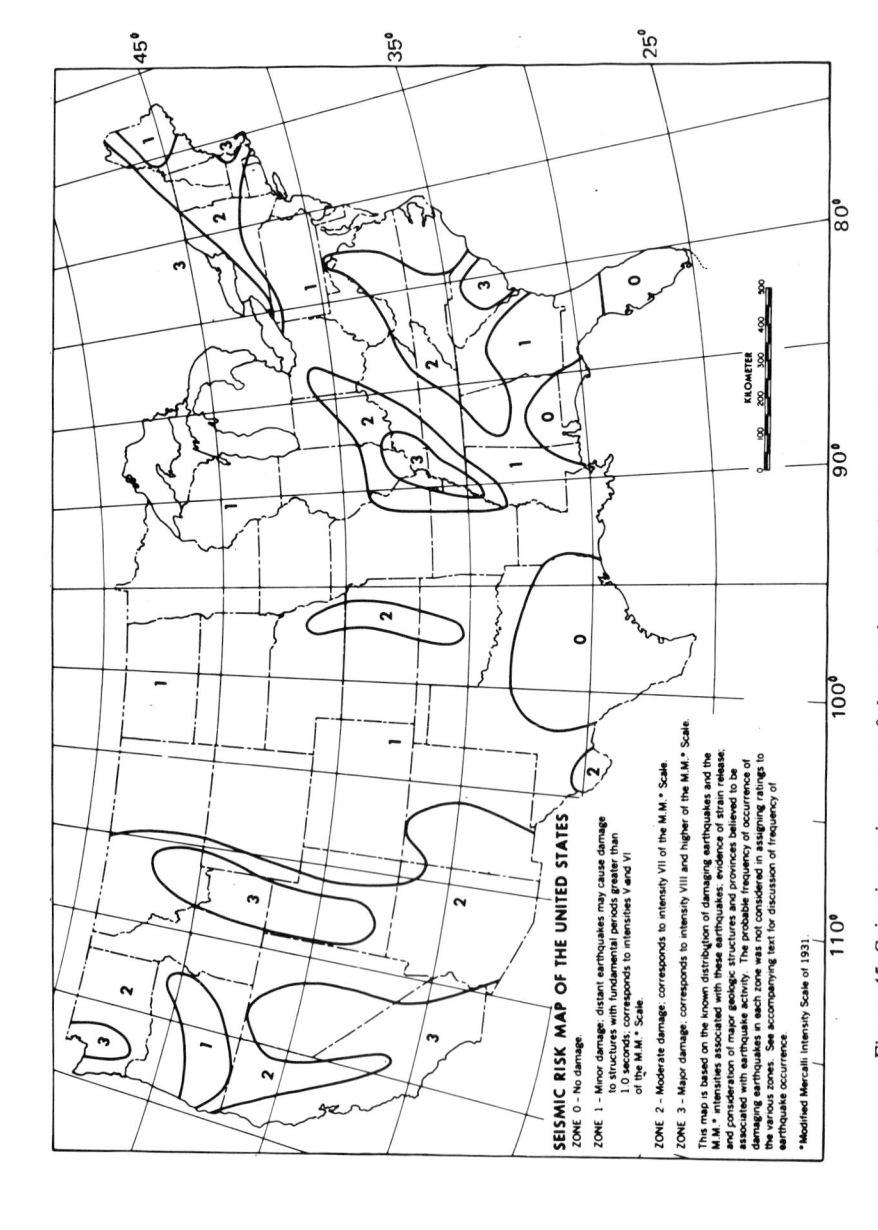

Figure 45. Seismic zoning map of the contiguous United States (Algermissen, Ref. 155).

Ulrich map. This map is based largely on the maximum Modified Mercalli intensity observed historically in each zone, but the spatial distribution of the intensities has been generalized to take into account some regional geological structures. The paper accompanying the zoning map also contained a maximum Modified Mercalli map, a strain energy release map, and earthquake recurrence curves for various regions of the country. The zoning map was adopted by the UBC in 1970, but the Code did not make use of the frequency of earthquake occurrence information that accompanied the map. In the 1976 edition of the UBC, the Algermissen map was modified to include a zone 4 in a portion of California, and in 1979 additional modifications were introduced for Idaho (Ref. 156 and Fig. 46). The introduction of the zone 4 in California had the effect qualitatively of taking into account the greater frequency of earthquakes of large magnitude that are possible in California.

The U.S. Departments of Housing and Urban Development (HUD) and Defense (DOD) have both used zoning maps similar to the UBC in their seismic design provisions for structures. HUD uses the Algermissen map in its *Minimum Property Standards for Single Family Dwellings* and *Minimum Property Standards for Multiple Family Housing* but has not adopted the zone 4 in California and other changes introduced by the UBC since 1976 (Refs. 157, 158).

Interest in the probabilistic estimation of ground motion increased in the 1960's as a result of the realization of the shortcomings of the existing hazard maps and because of the publication of a number of papers outlining possible probabilistic models and the application of these models to earthquake hazard estimation (for example, Refs. 159-164).

A probabilistic acceleration map for the contiguous United States was published by Algermissen and Perkins in 1976 (Ref. 169 and Fig. 47). The expected maximum acceleration in rock in a 50-year period with a 10 percent chance of being exceeded is the quantity mapped. The concept of hazard mapping used in the preparation of the map is that earthquakes are randomly distributed in magnitude, interoccurrence time, and space. The occurrence distribution in space is uniform within source zones.

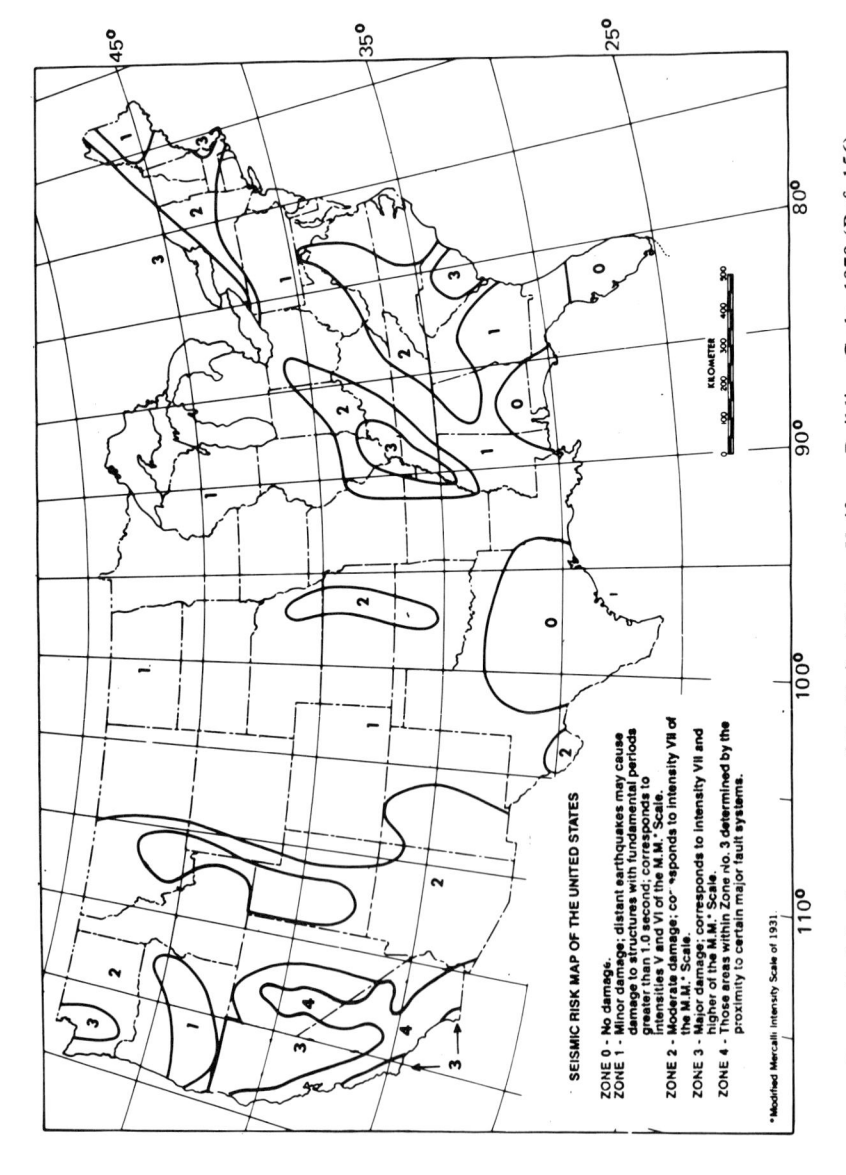

Figure 46. Seismic zone map of the United States, Uniform Building Code, 1979 (Ref. 156).

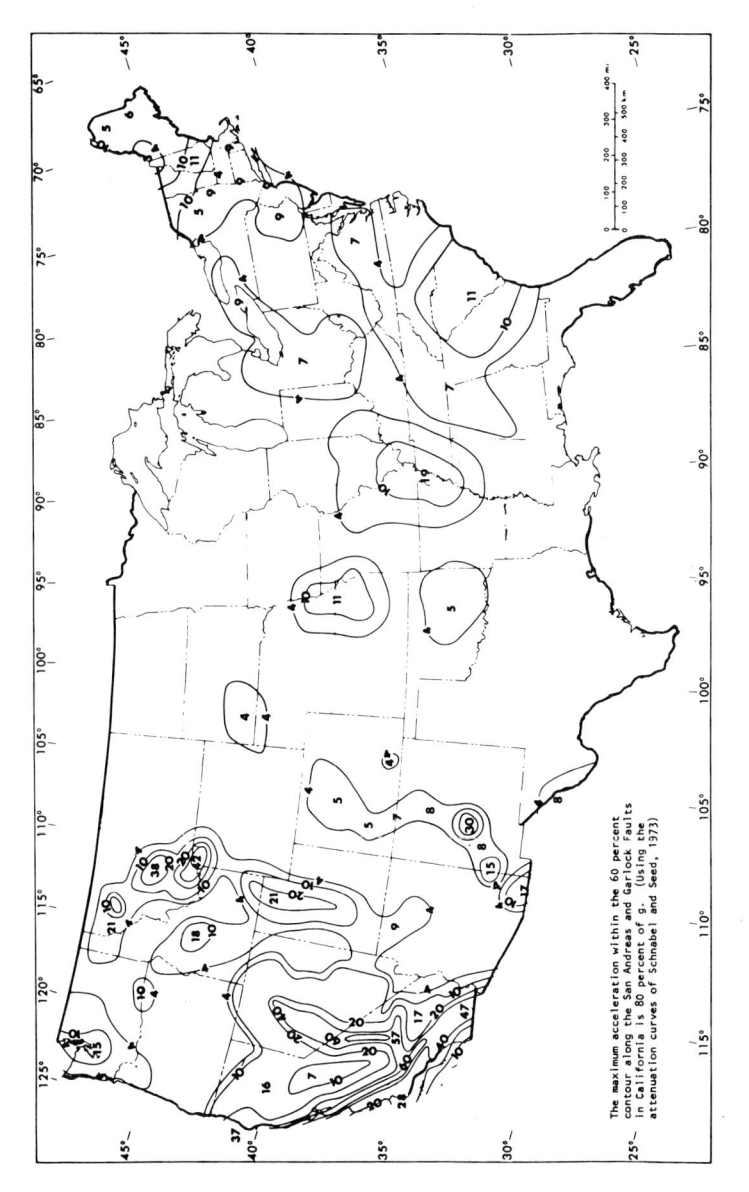

Figure 47. Probabilistic ground acceleration map of the conterminous United States, 50 year exposure time, 10 percent chance of exceedance, contours are percent of g (Algermissen and Perkins, 1976, Ref. 169).

Both the earthquake magnitudes and interoccurrence times have exponential distributions. Exponential interoccurrence times are characteristic of a Poisson process. The exponential magnitude distribution is an assumption based on empirical observation. The assumption of a Poisson process for earthquakes in time is consistent with historical earthquake occurrence insofar as it affects the probabilistic hazard calculation, provided the geographical areas considered are regional in nature. Large shocks closely approximate a Poisson process, but as magnitude decreases, earthquakes may depart significantly from this. However, ground motions associated with small earthquakes are of only marginal interest in engineering applications and consequently the Poisson assumption serves as a useful and simple model (Ref. 161). Spatially, the seismicity is modeled by grouping it into discrete areas termed "seismic source zones." The two general requirements for a seismic source zone are that (1) it has seismicity, and (2) it is a reasonable seismotectonic or seismogenic structure or zone. If a seismogenic structure or zone cannot be identified, the seismic source zone is based on historical seismicity. A seismotectonic structure or zone is taken to mean a specific geologic feature or group of features that is known to be associated with the occurrence of earthquakes. A seismogenic structure or zone is taken to mean a geologic feature or group of features for which the style of deformation and tectonic setting are similar and a relationship between this deformation and historic earthquake activity can be inferred.

The development of probabilistic ground motion maps depends on a knowledge of the attenuation of ground motion from the seismic sources to any site where the probabilistic ground motion is calculated. The acceleration attenuation curves used in the western United States for the development of the map shown in Fig. 47 are those published by Schnabel and Seed (Ref. 170). In the Midwest and East, the Schnabel and Seed curves were modified to reflect the lower rate of attenuation known to exist in these regions. The accelerations mapped are average maximum accelerations in material having a shear wave velocity of about 0.75 to 0.90 km/sec. Because of the dispersion in attenuation data and because local site conditions can greatly modify levels of ground shaking, regional and national hazard maps of the type prepared by Algermissen and Perkins are most useful as guides on a regional basis to expected ground motion and for comparison of the seismic hazard in various areas. For specific locations of interest, local site response and geological conditions should always be evaluated (Ref. 171). It is also useful to estimate the effect of parameter variability on the ground motion mapped. A number of interesting studies of the effects of parameter variability have been made (Refs. 165-168).

Completion of the probabilistic acceleration map of Algermissen and Perkins coincided with the developmental phase of a project undertaken by the Applied Technology Council (ATC) that had as its aim the development of new nationally applicable seismic design provisions. The results of the Applied Technology Council study were published in 1978 (Ref. 172).

The ATC report contains two ground motion maps based on *effective peak acceleration* and *effective peak velocity*, which are used to obtain "design ground shaking" and compute lateral force coefficients. For the conterminous United States, these two maps are based on the map of estimated acceleration in rock in a 50-year period at the 90-percent probability level developed by Algermissen and Perkins (1976). The Algermissen-Perkins map is also contained in the ATC report. The ATC Effective Peak Acceleration map (Fig. 48) is very similar to the Algermissen-Perkins acceleration map with the exception that the largest values of ground acceleration shown on the ATC map are 0.4 g in California, while the Algermissen-Perkins map has accelerations as high as 0.8 g in California. This implies that the probability of exceedance of 0.4 g is somewhat underestimated within the 0.4 g contours of the ATC map. The ATC Effective Peak Velocity map was derived from the Algermissen-Perkins acceleration map using principles and rules-of-thumb outlined in the report. Probabilistic maximum velocity and Modified Mercalli intensity maps of the United States prepared by Wiggins et al (Ref. 173) also provided important input for the preparation of the ground motion maps in the ATC report, particularly with regard to Alaska. The ATC-3 report is an excellent example of the use of

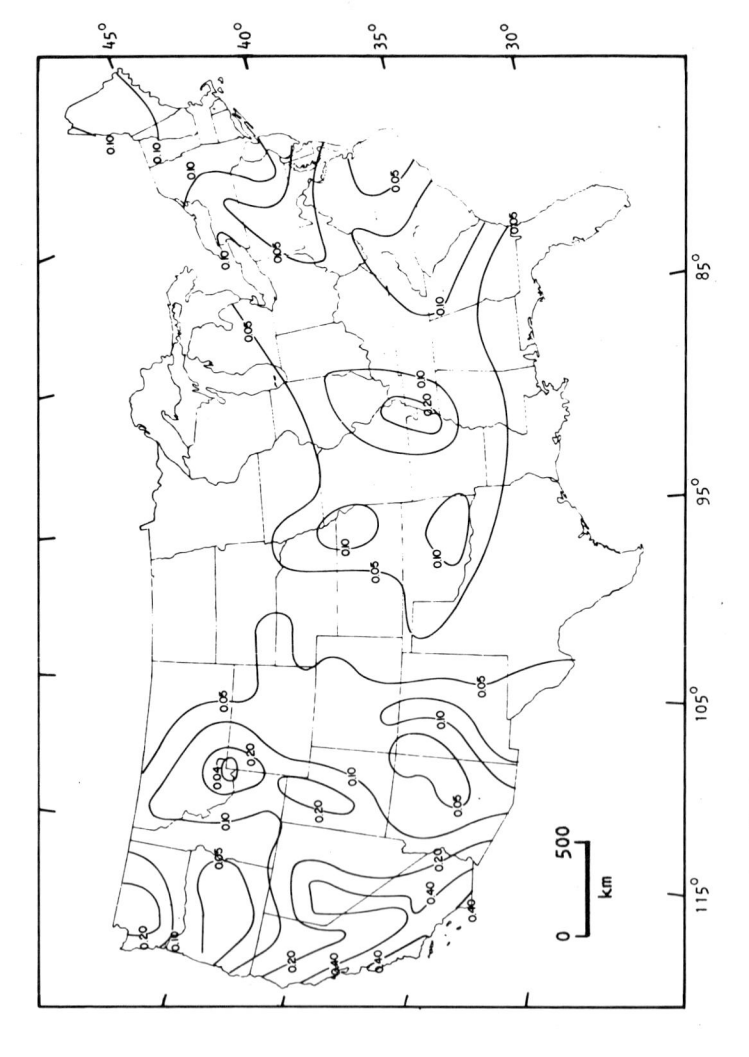

Figure 48. ATC effective peak acceleration map, Applied Technology Council, 1978 (Ref. 172).

recent research results in an interdisciplinary effort to produce new seismic design provisions.

A 50-year, 10 percent probability of exceedance acceleration map of Alaska and the adjacent offshore area was published by Thenhaus et al (Ref. 174 and Fig. 50a). However, Puerto Rico and the Virgin Islands have not as yet been given the attention they deserve in the geophysical and engineering literature, considering the level of known earthquake activity and tsunami occurrence in these areas.

In 1982, Algermissen et al (Ref. 175) published probabilistic maximum acceleration and velocity maps of the conterminous United States for exposure times (periods of interest) of 10, 50 and 250 years. Parameter variability is also extensively discussed in the report accompanying the maps. The 50-year, 10 percent chance of exceedance acceleration map of the contiguous United States is shown in Figs. 49 and 50 for comparison with the 1976 Algermissen-Perkins (Fig. 47). Considerable additional geological input was available for the delineation of seismic source zones used for the 1982 maps as compared with the source zones used for the 1976 map. This additional input resulted from a series of workshops held by the Geological Survey with invited regional experts from both within and outside the Survey. The most significant difference between the new 1982 ground motion maps and the 1976 Algermissen-Perkins map is the delineation of specific fault zones such as the Ramapo fault zone in New York-New Jersey and the Clarendon-Linden fault zone in northwestern New York (among others) as discrete seismic source zones. Often the net effect of this more detailed zoning on the resulting ground motion maps is to raise the ground motion in the vicinity of the geological structures identified as seismic source zones. Since ideas of the origins of seismicity, particularly in the eastern U.S. (Ref. 178), are rapidly changing as a result of research, ideas of and methods of delineating seismic source zones will also likely change in the future with a resulting change in the distributions of estimated ground motion.

Progress has also been made in the probabilistic evaluation of the potential for soil liquefaction (Ref. 176). Maps prepared to show the probability of liquefaction have so far been based on the

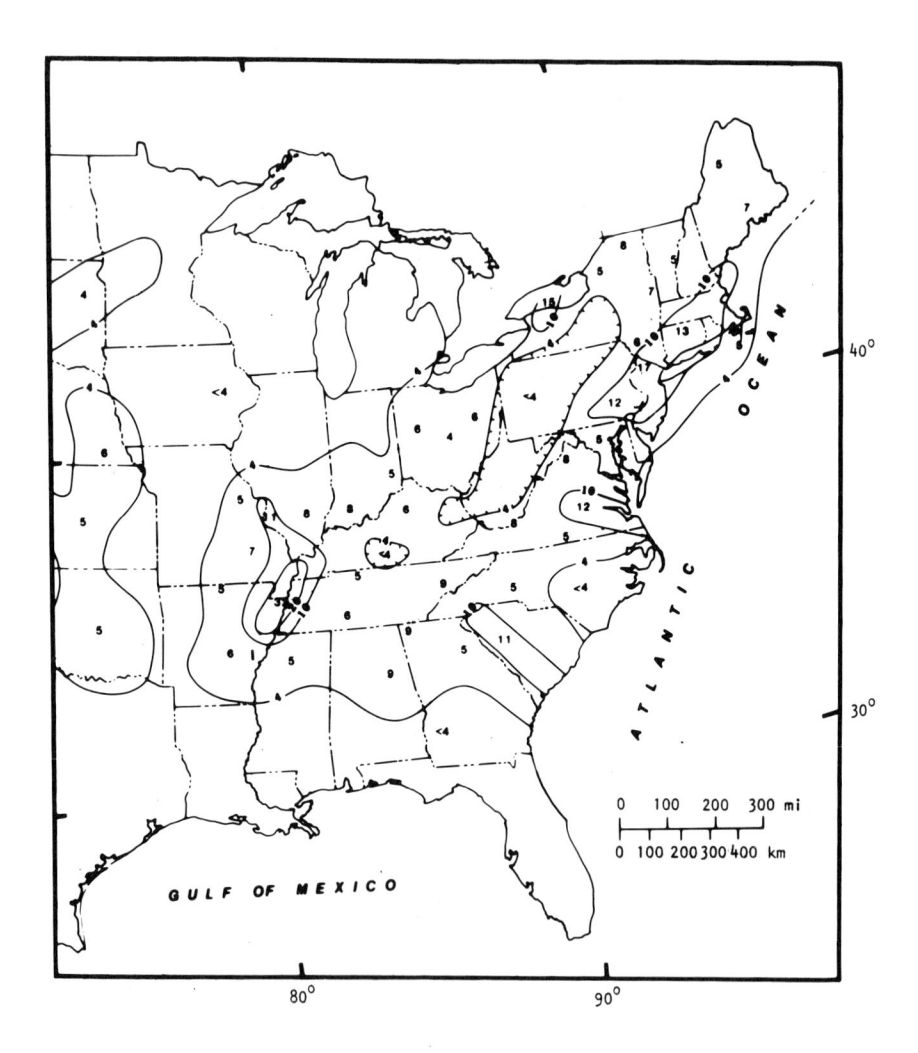

Figure 49. Probabilistic ground acceleration map of the eastern United States, 50 year exposure time, 10 percent chance of exceedance, contours are percent of g (Algermissen et al, 1982, Ref. 175).

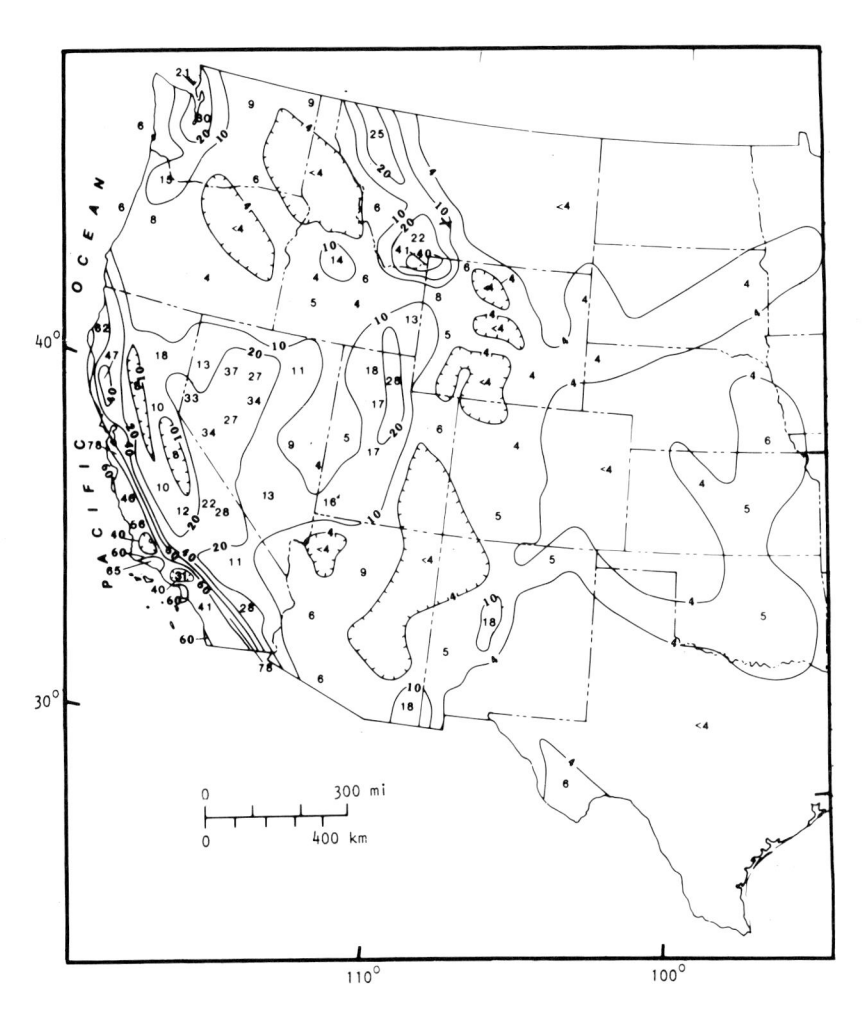

Figure 50. Probabilistic ground acceleration map of the western United States, 50 year exposure time, 10 percent chance of exceedance, contours are percent of g (Algermissen et al, 1982, Ref. 175).

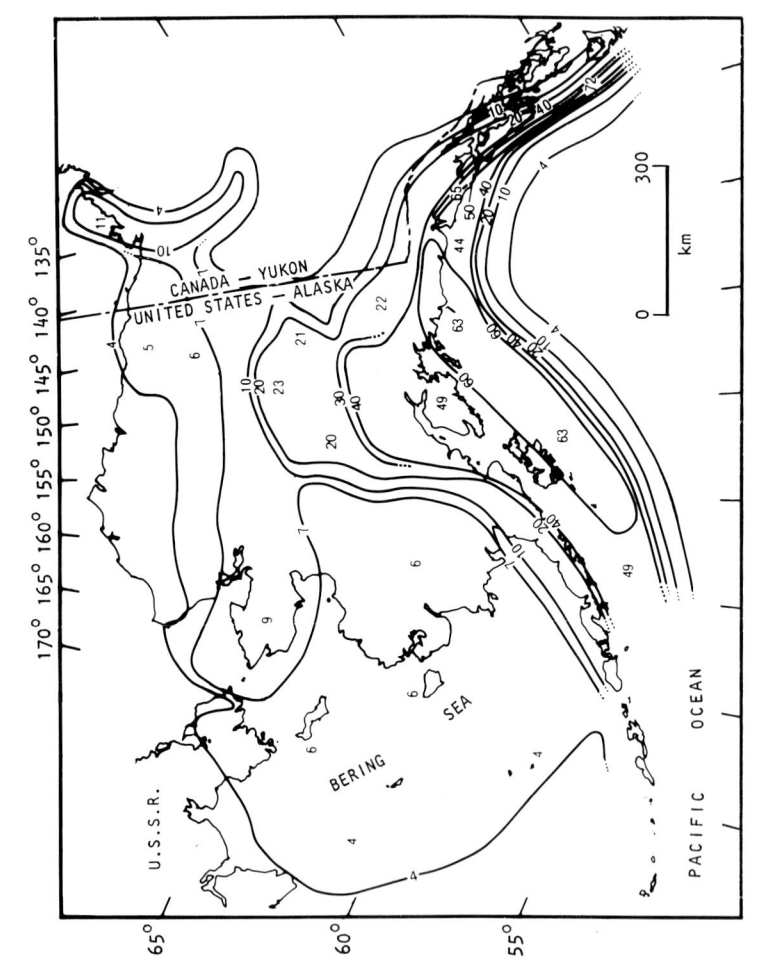

Figure 50a. Probabilistic ground acceleration map of Alaska, 50 year exposure time, 10 percent chance of exceedance, contours are percent of g (Thenhaus et al, 1982, Ref. 174).

same probabilistic ground motion models as were used in the development of the national ground motion maps already discussed. A general discussion of soil liquefaction may be found in another EERI monograph in this series, by Seed and Idriss (Ref. 177).

Improved ground motion attenuation relations, particularly for the eastern United States, Alaska, Hawaii, Puerto Rico and the Virgin Islands, are urgently needed. Research on the use of time-dependent probabilistic models for ground motion estimation is expanding, but it has been found that unless these time-dependent models are applied to specific geologic structures or faults where considerable data are available concerning the interoccurrence times of large earthquakes and unless the time since the last large event is known, time-dependent probabilistic models do not yield estimates of ground motion very different than those generated from a Poisson model.

ACKNOWLEDGEMENTS

I am greatly indebted to a number of my colleagues in the Geological Survey for suggestions regarding this Monograph. In particular, I should like to thank David Perkins for many useful suggestions and Paul Thenhaus, James Dewey and William Diment who reviewed the manuscript. I should also like to thank the EERI reviewers and in particular Dr. Anne Stevens, Earth Physics Branch, Department of Energy Mines and Resources, Ottawa, Canada for her many helpful suggestions. Lynn Barnhard and Bonny Askew provided invaluable assistance with the illustrations, and the manuscript was ably typed by Paula Ross. Joel Athey provided invaluable editorial assistance. Production graphics was handled by Gretchen Lillegraven.

APPENDIX A
MODIFIED MERCALLI INTENSITY SCALE

The intensity scale most commonly used in the United States since 1931 is the Modified Mercalli Intensity Scale. It is reproduced here in the original form as published (Ref. 1). Prior to 1931, the Rossi-Forel intensity scale was widely used in the United States. Since some important studies of early earthquakes use this scale, the corresponding Rossi-Forel intensity degree are shown at the corresponding degrees of the Modified Mercalli scale where the two scales agree. This comparison of the two scales also appeared in the original publication of the Modified Mercalli scale and it appears here as originally published. The original Modified Mercalli scale was later abridged, without complete success, since it contains some statements in conflict with the original scale (Ref. 2). The abridged version has, however, been widely used. See, for example, Ref. 15. A good discussion of the Modified Mercalli scale is provided by Richter (Ref. 2), who has attempted to quantify somewhat the kinds of structures maintained in the original scale. An interesting comparison of the Modified Mercalli Scale with other scales used throughout the world is provided in Ref. 185.

Modified Mercalli Intensity Scale of 1931

I

I
R.F.
Not felt — or, except rarely under especially favorable circumstances.
Under certain conditions, at and outside the boundary of the area in which a great shock is felt:
sometimes birds, animals, reported uneasy or disturbed;
sometimes dizziness or nausea experienced;
sometimes trees, structures, liquids, bodies of water, may sway — doors may swing, very slowly.

II

I   Felt indoors by few, especially on upper floors, or by
to   sensitive, or nervous persons.
II   Also, as in grade I, but often more noticeably:
R.F. sometimes hanging objects may swing, especially when delicately suspended;

sometimes trees, structures, liquids, bodies of water, may sway, doors may swing, very slowly;

sometimes birds, animals, reported uneasy or disturbed;

sometimes dizziness or nausea experienced.

III

III   Felt indoors by several, motion usually rapid vibration.
R.F. Sometimes not recognized to be an earthquake at first.

Duration estimated in some cases.

Vibration like that due to passing of light, or lightly loaded trucks, or heavy trucks some distance away.

Hanging objects may swing slightly.

Movements may be appreciable on upper levels of tall structures.

Rocked standing motor cars slightly.

IV

IV   Felt indoors by many, outdoors by few.
to   Awakened few, especially light sleepers.
V   Frightened no one, unless apprehensive from previous
R.F. experience.

Vibration like that due to passing of heavy, or heavily loaded trucks.

Sensation like heavy body striking building, or falling of heavy objects inside.

Rattling of dishes, windows, doors; glassware and crockery clink and clash.

Creaking of walls, frame, especially in the upper range of this grade.

Hanging objects swung, in numerous instances.
Disturbed liquids in open vessels slightly.
Rocked standing motor cars noticeably.

V

V	Felt indoors by practically all, outdoors by many or most; outdoors direction estimated.
to	
VI	Awakened many, or most.
R.F.	Frightened few — slight excitement, a few ran outdoors.

Building trembled throughout.
Broke dishes, glassware, to some extent.
Cracked windows — in some cases, but not generally.
Overturned vases, small or unstable objects, in many instances, with occasional fall.
Hanging objects, doors, swing generally or considerably.
Knocked pictures against walls, or swung them out of place.
Opened, or closed, doors, shutters, abruptly.
Pendulum clocks stopped, started, or ran fast, or slow.
Moved small objects, furnishings, the latter to slight extent.
Spilled liquids in small amounts from well-filled open containers.
Trees, bushes, shaken slightly.

VI

VI	Felt by all, indoors and outdoors.
to	Frightened many, excitement general, some alarm, many ran outdoors.
VII	
R.F.	Awakened all.

Persons made to move unsteadily.
Trees, bushes, shaken slightly to moderately.
Liquid set in strong motion.
Small bells rang — church, chapel, school, etc.
Damage slight in poorly built buildings.
Fall of plaster in small amount.

Cracked plaster somewhat, especially fine cracks, chimneys in some instances.
Broke dishes, glassware, in considerable quantity, also some windows.
Fall of knick-knacks, books, pictures.
Overturned furniture in many instances.
Moved furnishings of moderately heavy kind.

VII

VIII-R.F.
Frightened all — general alarm, all ran outdoors.
Some, or many, found it difficult to stand.
Noticed by persons driving motor cars.
Trees and bushes shaken moderately to strongly.
Waves on ponds, lakes, and running water.
Water turbid from mud stirred up.
Incaving to some extent of sand or gravel stream banks.
Rang large church bells, etc.
Suspended objects made to quiver.
Damage negligible in buildings of good design and construction, slight to moderate in well-built ordinary buildings, considerable in poorly built or badly designed buildings, adobe houses, old walls (especially where laid up without mortar), spires, etc.
Cracked chimneys to considerable extent, walls to some extent.
Fall of plaster in considerable to large amount, also some stucco.
Broke numerous windows, furniture to some extent.
Shook down loosened brickwork and tiles.
Broke weak chimneys at the roof-line (sometimes damaging roofs).
Fall of cornices from towers and high buildings.
Dislodged bricks and stones.
Overturned heavy furniture, with damage from breaking.
Damage considerable to concrete irrigation ditches.

VIII

VIII+   Fright general — alarm approaches panic.
to   Disturbed persons driving motor cars.
IX-   Trees shaken strongly — branches, trunks, broken off,
R.F.   especially palm trees.

Ejected sand and mud in small amounts.

Changes: temporary, permanent; in flow of springs and wells; dry wells renewed flow; in temperature of spring and well waters.

Damage slight in structures (brick) built especially to withstand earthquakes.

Considerable in ordinary substantial buildings, partial collapse: racked, tumbled down, wooden houses in some cases; threw out panel walls in frame structures, broke off decayed piling.

Fall of walls.

Cracked, broke, solid stone walls seriously.

Wet ground to some extent, also ground on steep slopes.

Twisting, fall, of chimneys, columns, monuments, also factory stacks, towers.

Moved conspicuously, overturned, very heavy furniture.

IX

IX+   Panic general.
R.F.   Cracked ground conspicuously.

Damage considerable in (masonry) structures built especially to withstand earthquakes:

    threw out of plumb some wood-frame houses built especially to withstand earthquakes;

great in substantial (masonry) buildings, some collapse in large part; or wholly shifted frame buildings off foundations, racked frames;

serious to reservoirs; underground pipes sometimes broken.

X

X
R.F.
Cracked ground, especially when loose and wet, up to widths of several inches; fissures up to a yard in width ran parallel to canal and stream banks.
Landslides considerable from river banks and steep coasts.
Shifted sand and mud horizontally on beaches and flat land.
Changed level of water in wells.
Threw water on banks of canals, lakes, rivers, etc.
Damage serious to dams, dikes, embankments.
Severe to well-built wooden structures and bridges, some destroyed.
Developed dangerous cracks in excellent brick walls.
Destroyed most masonry and frame structures, also their foundations.
Bent railroad rails slightly.
Tore apart, or crushed endwise, pipe lines buried in earth.
Open cracks and broad wavy folds in cement pavements and asphalt road surfaces.

XI

Disturbances in ground many and widespread, varying with ground material.
Broad fissures, earth slumps, and land slips in soft, wet ground.
Ejected water in large amount charged with sand and mud.
Caused sea-waves ("tidal") waves) of significant magnitude.
Damage severe to wood-frame structures, especially near shock centers.
Great to dams, dikes, embankments, often for long distances.
Few, if any (masonry), structures remained standing.

Destroyed large well-built bridges by the wrecking of supporting piers, or pillars.
Affected yielding wooden bridges less.
Bent railroad rails greatly, and thrust them endwise.
Put pipe lines buried in earth completely out of service

XII

Damage total — practically all works of construction damaged greatly or destroyed.
Disturbances in ground great and varied, numerous shearing cracks.
Landslides, falls of rock of significant character, slumping of river banks, etc., numerous and extensive.
Wrenched loose, tore off, large rock masses.
Fault slips in firm rock, with notable horizontal and vertical offset displacements.
Water channels, surface and underground, disturbed and modified greatly.
Dammed lakes, produced waterfalls, deflected rivers, etc.
Waves seen on ground surfaces (actually seen, probably, in some cases).
Distorted lines of sight and level.
Threw objects upward in the air.

APPENDIX B
MAGNITUDE

An excellent discussion of the various magnitude scales has been given by Nuttli (Ref. 179), and the following summary is largely taken from his paper. Magnitudes may be expressed in the form

$$M = \log_{10} A + G (r, h, f, \text{wave type})$$

where A is the amplitude of the ground motion at a distance r produced by a particular type of wave with frequency f. h is the focal depth and G takes into account the attenuation of the wave being considered as well as its source spectral amplitude. As pointed out by Nuttli, the various magnitudes are measures of the spectral amplitude of the source at discrete frequencies. M_L, the original magnitude developed by Richter (Ref. 2) for southern California, is an approximate measure of the 1 to 3 Hz portion of the source spectrum. The body wave magnitude m_b is a measure of the 1-Hz amplitude of the source spectrum, and the M_s (surface wave) magnitude is proportional to the source spectral amplitude of the 0.05 Hz, or 20 second period, waves. Nuttli has given a graphical relation between m_b and M_s for plate margin and mid plate earthquakes, and this relationship is shown in Figure 51. Nuttli (Ref. 180) has defined a magnitude used for classifying earthquakes in the central United States called m_{bLg} based on the sustained (third largest peak) amplitude of the crustal phase Lg recorded on a short period vertical seismograph used in the WWNSS system patterns. Herrmann and Nuttli have concluded (Ref. 181) that for magnitudes in the range from 3 to 5, m_{bLg} is essentially equivalent to M_L. The relationship has not been demonstrated for magnitudes greater than 5. In addition, they concluded that:

$$m_b \text{ (western U.S.)} = 0.99\, m_b \text{ (eastern U.S.)} - 0.39$$

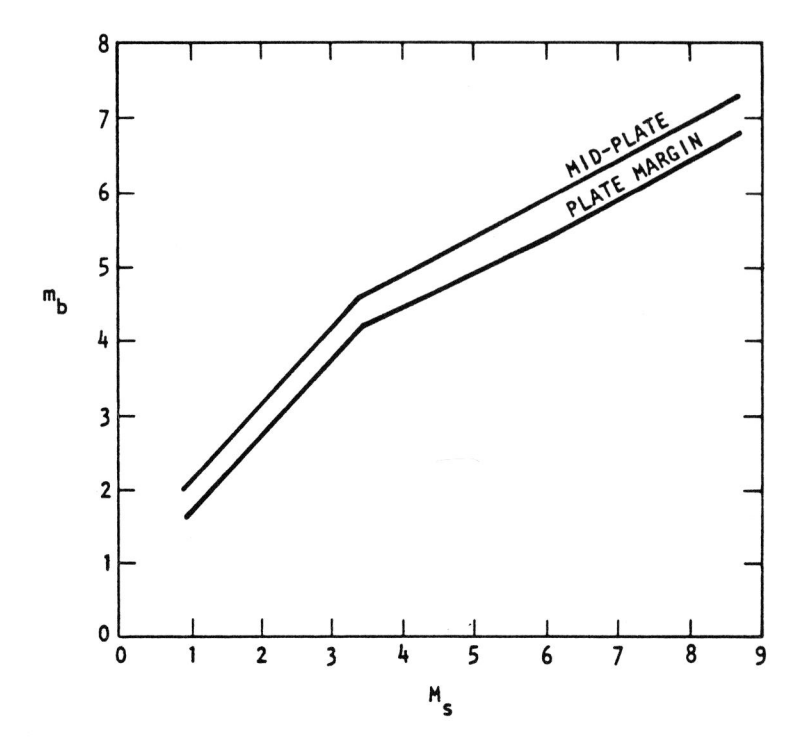

Figure 51. M_s - m_b relationships for plate margin and mid-plate earthquakes (after Ref. 179).

The difference in m_b for west and east is attributed to differences in upper mantle structure. Street and Turcotte (Ref. 182) give relationships between M_{bLg} and M_s for northeastern U.S. earthquakes.

In California, the magnitudes reported are usually M_L up to magnitudes of 6.5 to 6.7 and M_s for larger shocks. The explanation of the use of M_L outside California (for example, in Utah) is given in Ref. 72. For very large earthquakes, greater than about M_s = 8.6, all of the magnitude scales discussed are saturated (although not all of the scales saturate at the same magnitude level). For shocks larger than $M_s \simeq 8.6$, it is convenient to use M_w, the moment magnitude. M_w is defined so that it equals M_s for magnitudes at which M_s is not saturated.

The seismic moment is $M_o = \mu A d$, where μ is the rigidity modulus of the crust, A is the area of the ruptured fault surface and d is the average displacement of that surface.

The moment magnitude can be calculated from Mo by making assumptions about the stress drop in earthquakes and energy relationship developed by Gutenberg and Richter (Ref. 2). Kanamori (Ref. 183) gives the derivation of M_w and also lists great earthquakes for which M_w has been computed.

References

1. Wood, H. O. and F. Neumann, "Modified Mercalli Intensity Scale of 1931," *Bull. Seis. Soc. Amer.*, 21, 1931, pp. 277-283.

2. Richter, C. F., *Elementary Seismology*, W. H. Freeman, San Francisco, 1958.

3. O'Brien, L. J., *The Correlation of Response Spectral Amplitudes with Seismic Intensity*, U. S. Nuclear Regulatory Comm., NUREG/CR-1259, 1980.

4. Algermissen, S. T., "The Problem of Seismic Zoning," *Building Practices for Disaster Mitigation*, Building Science Series 46, U.S. Dept. of Commerce, National Bureau of Standards, 1973, pp. 112-125.

5. Smith, W. E. T., "Earthquakes of Eastern Canada and Adjacent Areas 1534-1927," *Pub. of Dominion Observatory, Ottawa*, 26:5, 1962, pp. 271-301.

6. Poppe, B. B., *Historical Survey of U. S. Seismograph Stations*, U. S. Geological Survey Prof. Paper 1096, 1979.

7. Bolt, B. A., "Seismicity of the Western United States," *Reviews in Engineering Geology, Vol. IV, Geology in the Siting of Nuclear Power Plants*, Geological Soc. of Amer., Boulder CO, 1979, pp. 95-107.

8. Stepp, J. C., W. A. Rinehart, and S. T. Algermissen, *Earthquakes in the United States 1963-64 and an Evaluation of the Detection Capability of the United States Seismograph Stations*, U. S. Dept. of Commerce, ESSA, Coast and Geodetic Survey, 1965.

9. Mauk, F. J. and D. H. Christensen, *A Probabilistic Evaluation of Earthquake Detection and Location Capability for Illinois, Indiana, Kentucky, Ohio, and West Virginia*, U. S. Nuclear Regulatory Comm., NUREG/CR-1648, 1980.

10. Oliver, J. and L. M. Murphy, "WWNSS: Seismology's Global Network of Observing Stations," *Science*, Vol. 174, 1971, pp. 254-261.

11. Peterson, J. and N. Orsini, "Seismic Research Observatories: Upgrading the Worldwide Seismic Data Network," Trans. Amer. Geophysical Union, *EOS*, 1976, pp. 548-556.

12. *Preliminary Determination of Epicenters* (pub. weekly), U. S. Geological Survey, National Earthquake Information Service

13. *Preliminary Determination of Epicenters Monthly Listing*, U. S. Dept. of Interior, U. S. Geological Survey, National Earthquake Information Service. (available by subscription)

14. *Earthquakes in the United States*, published quarterly as a U. S. Geological Survey Circular. Available from U. S. Geol. Survey, 604 S. Pickett, Alexandria, VA 22304.

15. *United States Earthquakes*, published annually since 1928. Currently published jointly by the U. S. Dept. of Interior, Geological Survey and the U. S. Dept. of Commerce, National Oceanic and Atmospheric Admin.

16. Le Pichon, X., J. Francheteau, and J. Bonnin, *Plate Tectonics*, Developments in Geotectonics 6, Elsevier Sci. Pub., 1973.

17. Sullivan, W., *Continents in Motion*, McGraw-Hill, New York, 1974.

18. Coffman, J. L., C. A. von Hake, and C. W. Stover (eds.), *Earthquake History of the United States*, U. S. Dept. of Commerce, NOAA and U. S. Dept. of Interior, Geological Survey, Pub. 41-1, Rev. Ed. (through 1970, with supplement through 1980), 1973.

19. Smith, W. E. T., "Earthquakes of Eastern Canada and Adjacent Areas, 1928-1959," *Pub. of Dominion Observatory*, Ottawa, 32:3, 1966, pp. 87-121.

20. Leblanc, G., "A Closer Look at the September 16, 1732, Montreal Earthquake," *Canada Jnl. Earth Sciences*, Vol. 18, 1981, pp. 539-550.

21. Stevens, A. E., ed., *Preliminary report of the Miramichi, New Brunswick, Canada, Earthquake Sequence of 1982*, EERI Special Report, 1983.

22. Basham, P. W., D. H. Weichert, and M. J. Berry, "Regional Assessment of Seismic Risk in Eastern Canada," *Bull. Seis. Soc. Amer.*, 69:5, Oct. 1979, pp. 1567-1602.

23. Woollard, G. P., "Tectonic Activity in North America as Indicated by Earthquakes," *The Earth's Crust and Upper Mantle*, ed. P. J. Hart, Geophys. Monograph 13, Trans. Am. Geophys. Union, Washington D C, pp. 125-133, 1969.

24. Barosh, P. J., "Cause of Seismicity in the Eastern United States, A Preliminary Appraisal," *Earthquakes and Earthquake Engineering*, Vol. 1, ed. J. E. Beavers, Ann Arbor Science Pub., pp. 397-417, 1981.

25. Sykes, L. R., "Intraplate Seismicity, Reactivation of Preexisting Zones of Weakness, Alkaline Magnetism, and Other Tectonism Postdating Continental Fragmentation," *Reviews of Geophysics and Space Physics*, Vol. 16, 1978, pp. 621-688.

26. Sbar, M. L. and L. R. Sykes, "Seismicity and Lithospheric Stress in New York and Adjacent Areas," *Jnl. Geophysical Research*, Vol. 82, 1977, pp. 5771-5786.

27. Diment, W. H., T. C. Urban, and F. A. Revetta, "Some Geophysical Anomalies in the Eastern United States," *Nature of the Solid Earth*, ed. E. Robertson, McGraw-Hill, New York, 1972, pp. 544-574.

28. Sbar, M. L. and L. R. Sykes, "Contemporary Compressive Stress and Seismicity in Eastern North America: an Example of Intraplate Tectonics," *Bull. Geological Soc. Amer.*, Vol. 84, 1973, pp. 1861-1882.

29. Yang, J.-P. and Y. P. Aggarwal, "Seismotectonics of Northeastern United States and Adjacent Canada," *Jnl. Geophysical Research*, Vol. 86, 1981, pp. 4981-4998.

30. Shakal, A. F. and M. N. Toksoz, "Earthquake Hazard in New England," *Science*, 1977, pp. 171-173.

31. Rankin, D. W., "Studies Related to the Charleston, South Carolina, Earthquake of 1886 — Introduction and Discussion," *Studies Related to the Charleston, South Carolina Earthquake of 1886 — A Preliminary Report*, ed. D. W. Rankin, U. S. Geological Survey Prof. Paper 1028, 1977, pp. 1-14.

32. Bollinger, G. A., "Seismicity of the Southeastern United States," *Bull. Seis. Soc. Amer.*, 63:5, Oct. 1973, pp. 1785-1808.

33. Bollinger, G. A., "Seismicity and Crustal Uplift in the Southeastern United States", *American Jnl. Science*, Vol. 273A, 1973, pp. 396-408.

34. Dutton, C. E., *The Charleston Earthquake of August 31, 1886*, U. S. Geological Survey 9th Annual Report 1887-89, pp. 203-528.

35. Armbruster, J. G. and L. Seeber, "Intraplate Seismicity in the Southeastern U. S. and the Appalachian Detachment," *Earthquakes and Earthquake Engineering: The Eastern United States*, Vol. 1, ed. J. E. Beavers, Ann Arbor Science Pub., 1981, pp. 376-395.

36. Tarr, A. C., "Recent Seismicity Near Charleston, South Carolina, and its Relationship to the August 31, 1886 Earthquake," *Studies Related to the Charleston, South Carolina Earthquake of 1886 — A Preliminary Report*, ed. D. W Rankin, U. S. Geological Survey Prof. Paper 1028, 1977, pp. 43-57.

37. Bollinger, G. A., "Reinterpretation of the Intensity Data for the 1886 Charleston, South Carolina Earthquake," *Studies Related to the Charleston, South Carolina Earthquake of 1886 — A Preliminary Report*, ed. D. W. Rankin, U. S. Geological Survey Prof. Paper 1028, 1977, pp. 17-32.

38. Nuttli, O. W., G. A. Bollinger, and D. W. Griffiths, "On the Relation Between Modified Mercalli Intensity and Body-Wave Magnitude," *Bull. Seis. Soc. Amer.*, 69:3, Jun. 1979, pp. 893-909.

39. Bollinger, G. A., "The Giles County, Virginia, Seismic Zone-Configuration and Hazard Assessment," *Earthquakes and Earthquake Engineering: The Eastern United States*, Vol. 1, ed. J. E. Beavers, Ann Arbor Science Pub., 1981, pp. 277-308.

40. Bollinger, G. A. and R. L. Wheeler, "The Giles County, Virginia Seismic Zone," *Science*, 219:4588, 1983, pp. 1063-1065.

41. Fletcher, J. B., M. L. Sbar, and L. B. Sykes, "Seismic Trends and Traveltime Residuals in Eastern North America and their Tectonic Implications," *Bull. Geological Soc. Amer.*, Vol. 89, 1978, pp. 1656-1676.

42. Hamilton, R. M., "Geologic Origin of Eastern Seismicity," *Earthquakes and Earthquake Engineering: The Eastern United States*, Vol. 1, ed. J. E. Beavers, Ann Arbor Science Pub., 1981, pp. 3-24.

43. Cook, F. A., D. S. Albaugh, L. D. Brown, S. Kaufmann, J. E. Oliver and R. D. Hatcher, "Thin-skinned Tectonics in the Crystalline Southern Appalachians: COCORP Seismic-Reflection Profiling of the Blue Ridge and Piedmont," *Geology*, Vol. 7, 1979, pp. 563-567.

44. Nuttli, O. W., "The Mississippi Valley Earthquakes of 1811 and 1812: Intensities, Ground Motion, and Magnitudes," *Bull. Seis. Soc. Amer.*, 63:1, Feb. 1973, pp. 227-248.

45. Nuttli, O., *Evaluation of Past Studies and the Identification of Needed Studies of the Effects of Major Earthquakes Occurring in the New Madrid Fault Zone*, Federal Emergency Management Agency, Kansas City, 1981, 28 pp. and appendices.

46. Nuttli, O. W., *The Mississippi Valley Earthquakes of 1811 and 1812*, Earthquake Information Bull., 6:2, U. S. Geological Survey, 1974, pp. 8-13.

47. Fuller, M. L., *The New Madrid Earthquake*, U. S. Geological Survey Bull. 494, 1912.

48. Nuttli, O. W., "Seismicity of the Central United States," *Geology in the Siting of Nuclear Power Plants*, eds. A. W. Hatheway and C. R. McClure, Geological Society of America, Reviews in Engineering Geology, Vol. 4, 1979, pp. 67-94.

49. Horner, R. B. and Hasegawa, H. S., "The Seismotectonics of Southern Saskatchewan," *Canadian Jnl. Earth Sciences*, Vol. 15, 1978, pp. 1341-1355.

50. Nuttli, O. W. and Herrmann, R. B., *Credible Earthquakes for the Central United States, State-of-the-Art for Assessing Earthquake Hazards in the United States*, Report No. 12, Misc. Papers S-73-1, U. S. Army Waterways Experiment Station, Vicksburg, MS, 1978.

51. Stauder, W. J., "Microearthquake Array Studies of the Seismicity in Southwest Missouri," *Earthquake Information Bull.*, 9:1, 1977, U. S. Geological Survey, pp. 8-13.

52. Stauder, W. J., R. B. Herrmann, S. Singh, R. Perry, E. Haug, and S. T. Morrissey, *Central Mississippi Valley Earthquake Bulletin*, Quarterly Report no. 21, 1979.

53. Hildenbrand, T. G., M. F. Kane, and J. D. Hendricks, *Magnetic Basement in the Upper Mississippi Embayment Region — A Preliminary Report*, U. S. Geological Survey Prof. Paper 1236-E, 1983.

54. Sawkins, F. J., "Widespread Continental Rifting: Some Considerations of Timing and Mechanism," *Geology*, Vol. 4, 1976, pp. 427-430.

55. Ervin, C. P. and L. D. McGinnis, "The Reelfoot Rift — Reactivated Precursor of the Mississippi Embayment," *Geological Soc. Amer. Bull.*, Vol. 86, 1975, pp. 1287-1295.

56. Hildenbrand, T. G., M. F. Kane, and W. J. Stauder, *Magnetic and Gravity Anomalies in the Northern Mississippi Embayment and their Spatial Relation to Seismicity*, U. S. Geological Survey Miscellaneous Field Studies Map MF-914, 1977.

57. Zoback, M. D., R. M. Hamilton, A. J. Crone, D. P.. Russ, F. A. McKeown, and S. R. Brockman, "Recurrent Intraplate Tectonism in the New Madrid Seismic Zone," *Science*, Vol. 209, 1980, pp. 971-976.

58. Zoback, M. L. and M. D. Zoback, "State of Stress in the Conterminous United States," *Jnl. Geophysical Research*, Vol. 85, 1980, pp. 6113-6156.

59. Hamilton, R. M., "Tectonic Evolution of the Northern Mississippi Embayment, Implications for New Madrid Seismicity," *Trans. Amer. Geophysical Union, EOS*, Vol. 61, 1980, p. 1026.

60. Heyl, A. V. and F. A. McKeown, *Preliminary Seismotectonic Map of Central Mississippi Valley and Environs*, U. S. Geological Survey Miscellaneous Field Studies Map MF-1011, 1978.

61. Bradley, E. A. and T. J. Bennett, "Earthquake History of Ohio," *Bull. Seis. Soc. Amer.*, 55:4, Aug. 1965, pp. 745-752.

62. Hinze, W. J., L. W. Braile, G. R. Keller, E. G. Lidiak, *A Tectonic Overview of the Central Midcontinent*, U. S. Nuclear Regulatory Comm., NUREG-0382, 1977.

63. Stauder, W. J. and O. W. Nuttli, "Seismic Studies: South Central Illinois Earthquake of November 9, 1968," *Bull. Seis. Soc. Amer.*, 60:3, Jun. 1970, pp. 973-981.

64. Merriam, D., "History of Earthquakes in Kansas," *Bull. Seis. Soc. Amer.*, 46:2, Apr. 1956, pp. 87-96.

65. DuBois, S. and F. W. Wilson, *A Revised and Augmented List of Earthquake Intensities for Kansas, 1867-1977*, U. S. Nuclear Regulatory Comm., NUREG/CR-0294, 1978.

66. Steeples, D., S. DuBois, and F. W. Wilson, "Seismicity, Faulting and Geophysical Anomalies in Nemaha County, Kansas: Relationship to Regional Structures," *Geology*, Vol. 7, 1979, pp. 134-138.

67. Herrmann, R. B., C. A. Langston, and J. E. Zollweg, "The Sharpsburg, Kentucky, Earthquake of 27 July, 1980," *Bull. Seis. Soc. Amer.*, 72:4, Aug. 1982, pp. 1219-1239.

68. Zollweg, J. E., "Aftershocks of the Sharpsburg, Kentucky Earthquake of July 27, 1980, A Preliminary Report," (abstract) *Earthquake Notes*, Vol. 51, 1980, p. 39.

69. Street, R., "Ground Motion Values Obtained for the 27 July, 1980 Sharpsburg, Kentucky, Earthquake," *Bull. Seis. Soc. Amer.*, 72:4, Aug. 1982, pp. 1295-1307.

70. Reagor, B. G., C. W. Stover, and M. G. Hopper, *Preliminary Report of the Distribution of Intensities for the Kentucky Earthquake of July 27, 1980*, U. S. Geological Survey Open-File Report 81-198, 1981.

71. Arabasz, W. J., R. B. Smith, and W. D. Richins, eds., *Earthquake Studies in Utah, 1850-1978*, Special Pub. Univ. of Utah Seismograph Stations, Dept. of Geology and Geophysics, 1979.

72. Griscom, M. and W. J. Arabasz, "Local Magnitude Region: Wood-Anderson Calibration, Coda-duration Estimates of M_L, and M_L versus m_b," *Earthquake Studies in Utah, 1850-1978*, eds. W. J. Arabasz, R. B. Smith and W. D. Richins, Special Pub. Univ. of Utah, 1979.

73. Smith, R. B., "Seismicity, Crustal Structure, and Intraplate Tectonics of the Interior of the Western Cordillera", *Cenozoic Tectonics and Regional Geophysics of the Western Cordillera*, eds. R. B. Smith and G. P Eaton, Geological Soc. Amer. Memoir 152, 1978, pp. 111-144.

74. Qamar, A. and B. Hawley, "Seismic Activity near the Three Forks Basin, Montana," *Bull. Seis. Soc. Amer.*, 69:6, Dec. 1979, pp. 1917-1929.

75. Arabasz, W. J., W. D. Richins, and C. J. Langer, "The Idaho-Utah (Pocatello Valley) Earthquake Sequence of March-April 1975," *Earthquake Studies in Utah, 1850-1978*, eds. W. J. Arabasz, R. B. Smith, and W. D. Richins, Special Pub. Univ. of Utah, 1979.

76. Rogers, A. M., C. J. Langer, and R. C. Bucknam, "The Pocatello Valley, Idaho, Earthquake," *Earthquake Information Bull.*, Vol. 7, 1975, U.S. Geological Survey, pp. 16-18.

77. Arabasz, W. J., R. B. Smith, and W. D. Richins, "Earthquake Studies along the Wasatch Front, Utah: Network Monitoring Seismicity and Seismic Hazards," *Proc. 10th Conf. Earthquake Hazards Along the Wasatch-Sierra Nevada Frontal Fault Zones*, U. S. Geological Survey Open-File Report 80-801, 1980.

78. Swan, F. H., D. P. Schwartz, and L. A. Cluff, "Recurrence of Moderate-to-Large Magnitude Earthquakes Produced by Surface Faulting on the Wasatch Fault Zone, Utah," *Bull. Seis. Soc. Amer.*, 70:5, Oct. 1980, pp. 1431-1462.

79. DuBois, S. and A. Smith, *The 1887 Earthquake in San Bernardino Valley, Sonora: Historic Accounts and Intensity Patterns in Arizona*, Spec. Paper No. 3, Bureau of Geology and Mineral Tech., Arizona, 1980.

80. DuBois, S., M. L. Sbar, and T. Nowak, *Historical Seismicity in Arizona*, U. S. Nuclear Regulatory Comm., NUREG CR-2577, 1982.

81. Eberhart-Phillips, D., R. M. Richardson, M. L. Sbar, and R. B. Herrmann, "Analysis of the 4 February 1976 Chino Valley, Arizona, Earthquake," *Bull. Seis. Soc. Amer.*, 71:3, Jun. 1981, pp. 787-801.

82. Northrop, S., "New Mexico's Earthquake History, 1849-1975" *Tectonic and Mineral Resources of Southwestern New Mexico*, eds. L. Woodward and S. Northrop, New Mexico Geological Survey, Special Pub. No. 6, 1976, pp. 77-87.

83. Cordell, L., "Regional Geophysical Setting of the Rio Grande Rift," *Bull. Geological Soc. Amer.*, Vol. 89, 1978, pp. 1073-1090.

84. Sanford, A. R., K. Olsen, and L. Jaksha, "Seismicity of the Rio Grande Rift," *Rio Grande Rift: Tectonics and Magmation*, ed. R. E. Riecker, American Geophysical Union, Washington DC, 1979.

85. Sanford, A. R., K. Olsen, and L. Jaksha, *Earthquakes in New Mexico, 1849-1977*, Circular 171, New Mexico Bureau of Mines and Mineral Resources, 1981.

86. Herrmann, R. B., J. W. Dewey, and S-K. Park, "The Dulce, New Mexico, Earthquake of 23 January 1966", *Bull. Seis. Soc. Amer.*, 70:6, Dec. 1980, pp. 2171-2183.

87. Hollister, J. and R. Wermer, eds., "Geophysical and Geological Studies of the Relationships between the Denver Earthquakes and the Rocky Mountain Arsenal Well, Part A," *Quart. Colo. School of Mines*, Vol. 63, 1968.

88. Smith, R. B. and G. P. Eaton, eds., *Cenozoic Tectonics and Regional Geophysics of the Western Cordillera*, Geological Soc. of America Memoir 152, 1978.

89. Real, C. R., T. R. Toppozada, and D. L. Parke, *Earthquake Catalog of California, January 1, 1900-December 31, 1974*, California Div. Mines and Geology, Special Pub. 52 and magnetic tape, 1978.

90. Ryall, F. D. and G. M. Smith, *Bulletin of the Seismological Laboratory for the period January 1, 1975 to December 31, 1979*, Univ. of Nevada, Reno, 1980.

91. Slemmons, D. B., A. E. Jones, and J. I. Gimlett, "Catalog of Nevada Earthquakes, 1852-1960," *Bull. Seis. Soc. Amer.*, 55:2, Apr. 1965, pp. 519-565.

92. Slemmons, D. B., *Faults and Earthquake Magnitude*, Misc. Papers S-73-1, Report 6, U. S. Army Engineer Waterways Experiment Station, Vicksburg, MS., 1977.

93. Hill, D. P., "Seismic Evidence for the Structure and Cenozoic Tectonics of the Pacific Coast States," *Cenozoic Tectonics and Regional Geophysics of the Western Cordillera*, eds. R. B. Smith and G. P. Eaton, Geological Soc. of America Memoir 152, 1978, pp 145-174.

94. VanWormer, J. D. and A. S. Ryall, "Sierra Nevada-Great Basin Boundary Zone: Earthquake Hazard Related to Structure, Active Tectonic Processes, and Anomalous Patterns of Earthquake Occurrence," *Bull. Seis. Soc. Amer.*, 70:5, Oct. 1980, pp. 1557-1572.

95. Gary, M., R. McAfee and C. L. Wolf, eds., *Glossary of Geology*, Amer. Geol. Inst., 1974, p. 61.

96. Fenneman, N. M., *Physiography of the Western United States*, McGraw-Hill, New York, 1931.

97. Fenneman, N. M., *Physical Divisions of the United States* (map), U. S. Geological Survey, scale 1:7,000,000, 1946.

98. Wallace, R. E., "Profiles and Ages of Young Fault Scarps, North-central Nevada," *Bull. Geological Soc. America*, Vol. 88, 1977, pp. 1267-1281.

99. Wallace, R. E., *Patterns of Faulting and Seismic Gaps in the Great Basin Province*, U. S. Geological Survey Open-File Report 78-943, 1978.

100. Wallace, R. E., "Earthquake Recurrence Intervals on the San Andreas Fault," *Bull. Geological Soc. America*, Vol. 81, 1970, pp. 2875-2890.

101. Wallace, R. E., "Behavior of Different Segments of the San Andreas Fault," *Earthquake Information Bull.*, Vol. 10, 1978, U.S. Geological Survey, pp. 126-130.

102. Lawson, A. C. et al, *The California Earthquake of April 18, 1906*, Report of the State Earthquake Commission, 2 vols., and atlas, Washington DC, Carnegie Institution, 1908.

103. "The Fort Tejon, California Earthquake of 1857," *Earthquake Information Bull.*, Vol. 13, 1981, U.S. Geological Survey, pp. 85-87.

104. Agnew, D. C. and K. E. Sieh, "A Documentary Study of the Felt Effects of the Great California Earthquake of 1857," *Bull. Seis. Soc. Amer.*, 68:6, Dec. 1978, pp. 1717-1729.

105. Sieh, K. E. "Prehistoric Large Earthquakes Produced by Slip on the San Andreas Fault at Pallet Creek California," *Jnl. Geophysical Research*, Vol. 83, 1978, pp. 3907-3939.

106. Sieh, K. E., "Central California Foreshocks of the Great 1857 Earthquake," *Bull. Seis. Soc. Amer.*, 68:6, Dec. 1978, pp. 1731-1749.

107. Allen, C. R. and J. M. Nordquist, *Foreshock, Main Shock and Larger Aftershocks of the Borrego Mountain Earthquake*, U. S. Geological Survey Prof. Paper 787, 1972.

108. *The Imperial Valley Earthquake of October 15, 1979*, ed. G. S. Gohn, U. S. Geological Survey Prof. Paper 1254, 1982.

109. Allen, C. R., P. St. Amand, C. F. Richter, and J. M. Nordquist, "Relationship between Seismicity and Geologic Structure in the Southern California Region," *Bull. Seis. Soc. Amer.*, 55:4, Aug. 1965, pp. 753-798.

110. McKenzie, D. P. and R. L. Parker, "The North Pacific — An Example of Tectonics on a Sphere," *Nature*, Vol. 216, 1979, pp. 1276-1280.

111. *A Study of Earthquake Losses in the Puget Sound, Washington Area*, U. S. Geological Survey Open-File Report 75-375, 1975.

112. Rogers, G. C. and H. S. Hasegawa, "A Second Look at the British Columbia Earthquake of June 23, 1946," *Bull. Seis. Soc. Amer.*, 68:3, June 1978, pp. 653-675.

113. Riddihough, R. P., "A Model for Recent Plate Interactions off Canada's West Coast," *Canadian Jnl. Earth Science*, Vol. 14, 1977, pp. 384-396.

114. Riddihough, R. P., "The Juan de Fuca Plate," *Trans. Amer. Geophysical Union, EOS*, 59:9, 1978, pp. 836-842.

115. Riddihough, R. P. and R. D. Hyndman, "Canada's Active Western Margin — The Case for Subduction," *Geoscience Canada*, 3:4, 1977, pp. 269-278.

116. Kulm, L. D. and G. A. Fowler, "Oregon Continental Margin Structure and Stratigraphy — A Test of the Imbrecate Thrust Model," *The Geology of Continental Margins*, eds. C. A. Burke and C. L. Drake, Springer-Verlag, New York, 1974, pp. 261-291.

117. Atwater, T., "Implications of Plate Tectonics for the Cenozoic Tectonic Evolution of Western North America," *Bull. Geological Soc. Amer.*, Vol. 81, 1970, pp. 3513-3535.

118. Crosson, R. S., "Small Earthquakes, Structure and Tectonics of the Puget Sound Region," *Bull. Seis. Soc. Amer.*, 62:5, Oct. 1972, pp. 1133-1171.

119. Stacey, R. A., "Gravity Anomalies, Crustal Structure, and Plate Tectonics in the Canadian Cordillera," *Canadian Jnl. Earth Science*, Vol. 10, 1973, pp. 615-628.

120. Algermissen, S. T. and S. T. Harding, "Preliminary Seismological Report," *The Puget Sound, Washington Earthquake of April 29, 1965*, U. S. Dept. of Commerce, Coast and Geodetic Survey, 1965.

121. Wyss, M. and P. Molnar, "Source Parameters of Intermediate and Deep Focus Earthquakes in the Trega Arc," *Physical Earth Plane. Int.*, Vol. 6, 1972, pp. 279-292.

122. Milne, W. G., G. C. Rogers, R. P. Riddihough, G. A. McMechan, and R. D. Hyndman, "Seismicity of Western Canada," *Canadian Jnl. Earth Sciences*, Vol. 15, 1978, pp. 1170-1193.

123. Basham, P. W., D. H. Weichert, F. M. Anglin, and M. J. Berry, *New Probabilistic Ground Motion Maps of Canada: A Compilation of Earthquake Source Zones, Methods and Results*, Earth Physics Branch Open-File Report 82-33, Ottawa, 1982.

124. Scott, N. H., *Evaluation of the Epicenter and Intensity of the Pacific Northwest Earthquake of December 14, 1872*, prepared for Bechtel, Inc., San Francisco, Sep. 1976.

125. Woodward-Clyde Consultants, *Review of the Pacific Northwest Earthquake of December 14, 1872*, prepared for Washington Public Power Supply System, Dec. 1976.

126. Weston Geophysical Research, Inc., *Presentation of Significant Data and Conclusions concerning the 1872 Earthquake*, prepared for Washington Public Power Supply System, Dec. 1976.

127. Algermissen, S. T., R. J. Brazee, C. W. Stover, and L. C. Pakiser, *Maximum Intensity of the Washington Earthquake of December 14, 1872*, Report of USGS/NOAA Ad Hoc Working Group on Intensities of Historic Earthquakes, 1977.

128. Coombs, H. A., W. G. Milne, O. W. Nuttli, and D. B. Slemmons, *Report of the Review Panel on the December 14, 1872 Earthquake*, prepared for the Utilities of the Pacific Northwest, Dec. 1976.

129. Hopper, M. G., S. T. Algermissen, D. M. Perkins, S. R. Brockman, and E. P. Arnold, "The Earthquake of December 14, 1872, in the Pacific Northwest,"(abstract), Annual Meeting of the Seis. Soc. of Amer., 1982.

130. Gower, H. D., *Tectonic Map of the Puget Sound Region, Washington*, U. S. Geological Survey Open-File Report 78-426, 1978.

131. Skehan, J. W., *A Continental-Oceanic Crustal Boundary in the Pacific Northwest*, Bedford, MA, Air Force Cambridge Research Laboratories, Office of Aerospace Research, Scientific Report 3, 1965.

132. Lawrence, D. L., "Strike-slip Faulting Terminates the Basin and Range Province in Oregon," *Bull. Geological Soc. Amer.*, Vol. 87, 1976, pp. 846-850.

133. Page, R. A., "Estimating Earthquake Potential," *Earthquake Information Bull.*, Vol. 12, 1980, U.S. Geological Survey, pp. 17-24.

134. Davies, J., L. Sykes, L. House, and K. Jacob, "Shumagin Seismic Gap, Alaska Peninsula: History of Great Earthquakes, Tectonic Setting and Evidence for High Seismic Potential," *Jnl. Geophysical Research,* Vol. 86, 1981, pp. 3821-3855.

135. Perez, O. J. and K. H. Jacob, "Tectonic Model and Seismic Potential of the Eastern Gulf of Alaska and Yakataga Seismic Gap," *Jnl. Geophysical Research,* Vol. 85, 1980, pp. 7132-7150.

136. Tarr, A. C., "Global Seismicity Studies, Projections of the Hyperspace," *Earthquake Information Bull.*, Vol. 4, 1972, U.S. Geological Survey, pp. 4-11.

137. Sherburne, R. W., S. T. Algermissen, and S. T. Harding, "The Hypocenter, Origin Time and Magnitude of the Prince William Sound Earthquake of March 28, 1964," *The Prince William Sound, Alaska, Earthquake of 1964 and Aftershocks*, ed. F. J. Wood, U. S. Dept. of Commerce Pub. 10-3, Coast and Geodetic Survey, Vol. 2, 1969, pp. 49-67.

138. Algermissen, S. T., W. A. Rinehart, R. W. Sherburne, and W. H. Dillinger, "Preshocks and Aftershocks," *The Great Alaska Earthquake of 1964, Seismology and Geology*, National Acad. of Sciences, 1971, pp. 313-364.

139. Plafker, G., "Tectonics," *The Great Alaska Earthquake of 1964, Geology, Part A*, National Acad. of Sciences, 1971, pp. 46-122.

140. *The Great Alaska Earthquake of 1964*, 8 vols., Committee on the Alaska Earthquake of Div. of Earth Sciences, National Research Council, National Acad. of Sciences, Washington DC, 1971-72.

141. U. S. Geological Survey Professional Papers 541, 542, 543, 544, 545, 546, 1967-1970.

142. Wood, F. J., ed., *The Prince William Sound, Alaska Earthquake of 1964 and Aftershocks*, U. S. Dept. of Commerce, ESSA, Coast and Geodetic Survey, Pub. 10-3, 3 vols., 1969.

143. Wyss, M., F. W. Klein, and A. C. Johnston, "Precursors to the Kalapana M = 7.2 Earthquake," *Jnl. Geophysical Research*, Vol. 86, 1981, pp. 3881-3900.

144. Ando, M., "The Hawaii Earthquake of November 29, 1975: Low Dip Angle Faulting due to Forceful Injection of Magma," *Jnl. Geophysical Research*, Vol. 84, 1979.

145. Asencio, E., *Western Puerto Rico Seismicity*, U. S. Geological Survey Open-File Report 80-192, 1980.

146. U. S. Dept. of the Interior, Geological Survey, Progress Report, Third Quarter Fiscal Year 1977 to Puerto Rico Water Resources Authority, 1977.

147. Reid, H. F. and S. Taber, *The Puerto Rico Earthquake of 1918 — With Description of Earlier Earthquakes*, Report of the Earthquake Investigation Commission, U. S. House of Representatives, Document No. 269, 66th Congress, First Session, 1919.

148. Espinosa, A. F., ed., *The Guatemalan Earthquake of February 4, 1976, A Preliminary Report*, U. S. Geological Survey Prof. Paper 1002, 1976.

149. Jordan, T. H., "The Present-day Motions of the Caribbean Plate," *Jnl. Geophysical Research*, Vol. 80, 1975, pp. 4433-4439.

150. Freeman, J. R., *Earthquake Damage and Earthquake Insurance*, McGraw-Hill, New York, 1932.

151. Perkins, D. M., "Seismic Risk Maps," *Earthquake Information Bull.*, Vol. 6, 1974, U.S. Geological Survey, pp. 10-15.

152. Roberts, E. B. and F. P. Ulrich, "Seismological Activities of the U. S. Coast and Geodetic Survey in 1949," *Bull. Seis. Soc. Amer.*, 41:3, Jul. 1951, pp. 205-220.

153. Richter, C. F., "Seismic Regionalization," *Bull. Seis. Soc. Amer.*, 49:2, Apr. 1959, pp. 123-162.

154. *Uniform Building Code*, International Conference of Building Officials, Whittier, CA, 1970 (updated every 3 years).

155. Algermissen, S. T., "Seismic Risk Studies in the United States," *Proc. 4th World Conf. on Earthquake Engineering*, Santiago, Chile, 1969, Vol. 1, pp. 14-27.

156. *Uniform Building Code*, International Conference of Building Officials, Whittier, CA, 1979 (updated every 3 years).

157. Dept. of Housing and Urban Development, *Minimum Property Standards for One and Two Family Dwellings*, HUD Pub. 4900.1, 1982.

158. Dept. of Housing and Urban Development, *Minimum Property Standards for Multiple Family Housing*, HUD Pub. 4910.1, 1973.

159. Lomnitz, C., "Statistical Prediction of Earthquakes," *Research Geophysics*, Vol. 4, 1966, pp. 377-393.

160. Lomnitz, C, "An Earthquake Risk Map of Chile," *Proc. 4th World Conf. on Earthquake Engineering*, Santiago, Chile, 1969, Vol. 1, pp. 161-171.

161. Cornell, C. A., "Engineering Seismic Risk Analysis," *Bull. Seis. Soc. Amer.*, 58:5, Oct. 1968, pp. 1583-1606.

162. Estewa, L., "Seismicity Prediction: A Bayesian Approach," *Proc. 4th World Conf. on Earthquake Engineering*, Santiago, Chile, 1969, Vol. 1, pp. 172-184.

163. Cornell, C. A. and E. H. Vanmarcke, "The Major Influences on Seismic Risk," *Proc. 4th World Conf. on Earthquake Engineering*, Santiago, Chile, 1969, Vol. 1, pp. 69-84.

164. Milne, W. G. and A. G. Davenport, "Earthquake Probability," *Proc. 4th World Conf. on Earthquake Engineering*, Santiago, Chile, 1969, Vol. 1, pp. 55-68.

165. Esteva, L., "Seismic Risk and Seismic Design Decisions," *Seismic Design for Nuclear Power Plants*, ed. R. J. Hanson, MIT Press, Cambridge, 1970, pp. 142-182.

166. Donovan, N. C. and A. E. Bornstein, "Uncertainties in Seismic Risk Procedures," *Jnl. Geotech. Eng. Div., ASCE*, 104:GT7, Jul. 1978, pp. 869-887.

167. Perkins, D. M., "Effect of Changing Return Periods on Probabilistic Ground Motion," *Proc. Workshop on Seismic Performance of Underground Facilities*, Augusta, GA., E. I. du Pont de Nemours and Co., prepared for the Dept. of Defense, 1981, pp. 183-205.

168. McGuire, R. K. and K. M. Shedlock, "Statistical Uncertainties in Seismic Hazard Evaluations in the United States," *Bull. Seis. Soc. Amer.*, 71:4, Aug 1981, pp. 1287-1308.

169. Algermissen, S. T. and D. M. Perkins, *A Probabilistic Estimate of Maximum Acceleration in Rock in the Contiguous United States*, U. S. Geological Survey Open-File Report 76-416, 1976.

170. Schnabel, P. B. and H. B. Seed, "Accelerations in Rock for Earthquakes in the Western United States," *Bull. Seis. Soc. Amer.*, 63:2, Apr 1973, pp. 501-516.

171. Hays, W. W., *Procedures for Estimating Earthquake Ground Motions*, U. S. Geological Survey Prof. Paper 1114, 1980.

172. Applied Technology Council, *Tentative Provisions for the Development of Seismic Regulations for Buildings*, ATC 3-06, NBS Spec. Pub. 510, NSF Pub. 78-8, 1978.

173. Wiggins, J. H., J. L. Hirshberg, and A. Bronowicki, *Budgeting Justification for Earthquake Engineering Research*, Technical Report No. 75-1201-1, National Science Foundation, 1974.

174. Thenhaus, P. C., J. I. Ziony, W. H. Diment, M. G. Hopper, D. M. Perkins, S. L. Hanson, and S. T. Algermissen, "Probabilistic Estimates of Maximum Seismic Horizontal Ground Motion on Rock in Alaska and the Adjacent Outer Continental Shelf," *Alaska: Accomplishments during 1980*, ed. W. L. Coonrad, U. S. Geological Survey Circular 844, 1982.

175. Algermissen, S. T., D. M. Perkins, P. C. Thenhaus, S. L. Hanson, and B. L. Bender, *Probabilistic Estimates of Maximum Acceleration and Velocity in Rock in the Contiguous United States*, U. S. Geological Survey Open-File Report 82-1033, 1982.

176. Youd, T. L. and D. M. Perkins, "Mapping Liquefaction-Induced Ground Failure Potential," *Jnl. Geotech. Eng. Div., ASCE*, 104:GT4, Jul. 1978, pp. 433-446.

177. Seed, H. B. and I. M. Idriss, *Ground Motions and Soil Liquefaction During Earthquakes*, Earthquake Engineering Research Institute, Berkeley, CA, 1982.

178. Barstow, N. L., K. G. Brill, O. W. Nuttli, and P. W. Pomeroy, *An Approach to Seismic Zonation for Siting Nuclear Electric Power Generating Facilities in the Eastern United States*, U. S. Nuclear Regulatory Comm., NUREG/CR-1577.

179. Nuttli, O. W., "Similarities and Differences between Western and Eastern United States Earthquakes and their Consequences for Earthquake Engineering," *Earthquakes and Earthquake Engineering: Earthquakes in the Eastern United States*, Vol. 1, ed. J. E. Beavers, 1981, pp. 25-52.

180. Nuttli, O. W. "Seismic Wave Attenuation and Magnitude Relationship for Eastern North America," *Jnl. Geophysical Research*, Vol. 78, 1973, pp. 876-885.

181. Herrmann, R. B. and O. W. Nuttli, "Magnitude: The Relation of M_L to M_{bLg}," *Bull. Seis. Soc. Amer.*, 72:2, Apr. 1982, pp. 389-398.

182. Street, R. L. and F. T. Turcotte, "A Study of Northeastern North America Spectral Moments, Magnitudes, and Intensities," *Bull. Seis. Soc. Amer.*, 67:3, Jun. 1977, pp. 599-614.

183. Kanamori, H., "The Energy Release in Great Earthquakes," *Jnl. Geophysical Research*, Vol. 82, 1977, pp. 2981-2987.

184. Stover, C. W., B. G. Reagor, and R. J. Wetmiller, "Intensities and Isoseismal Map for the St. Elias Earthquake of February 28, 1979," *Bull. Seis. Soc. Amer.*, 70:5, Oct. 1980, pp. 1635-1649.

185. Barosh, P. J., *Use of Seismic Intensity Data to Predict the Effects of Earthquakes and Underground Nuclear Explosions in Various Geologic Settings*, U. S. Geological Survey Bulletin 1279, 1969.